Better Homes and Gardens®

STEP-BY-STEP
BASIC
PLUMBING

BETTER HOMES AND GARDENS® BOOKS

Editor: Gerald M. Knox
Art Director: Ernest Shelton

Building and Remodeling Editor: Joan McCloskey
Building Books Editor: Larry Clayton
Building Books Associate Editor: Jim Harrold

Associate Art Directors: Neoma Alt West,
    Randall Yontz
Copy and Production Editors: David Kirchner,
    Lamont Olson, David A. Walsh
Assistant Art Director: Harijs Priekulis
Senior Graphic Designer: Faith Berven
Graphic Designers: Alisann Dixon, Linda Ford,
    Lyne Neymeyer, Lynda Haupert, Tom Wegner

Editor in Chief: Neil Kuehnl
Group Editorial Services Director: Duane Gregg
Executive Art Director: William J. Yates

General Manager: Fred Stines
Director of Publishing: Robert Nelson
Director of Retail Marketing: Jamie Martin
Director of Direct Marketing: Arthur Heydendael

Step-By-Step Basic Plumbing

Editors: Larry Clayton, Jim Harrold
Copy and Production Editor: David A. Walsh
Graphic Designer: Thomas Wegner
Contributing Writer: James A. Hufnagel
Technical Consultants: James Downing, Kenn Spahr
Drawings: Carson Ode

Acknowledgments

Our appreciation goes to the following companies
and associations for their help in the preparation of this book:
American Standard Inc.
Brass® Craft Manufacturing Co.
Delta Faucet Company, Division of Masco Corporation of Indiana
Elkay® Manufacturing Company
Genova, Inc.
Chas. W. Goering, Plumbing-Heating-Cooling
In-Sink-Erator Division, Emerson Electric Co.
International Association of Plumbing
    and Mechanical Officials
Moen®, a Division of Stanadyne
Nibco Inc.
Ondine™, a Division of Interbath, Inc.
P & M Manufacturing Company
Plumb Crazy (Discount) Store
Price Pfister

CONTENTS

INTRODUCTION

Probably you've always been one of those "leave plumbing to the plumbers" kind of people. Most, in fact, go that route whenever they need plumbing-repair work or a plumbing improvement in their home.

So why your sudden interest in learning how to be your own plumber? Perhaps you have a repair or two to make and can't seem to find anyone willing to do the work for a reasonable price. Or maybe you've just received a higher-than-expected bid on a bathroom or kitchen remodeling. Whatever your motivation, the simple fact is that today you pay "the price" for what you can't do yourself.

The idea of doing plumbing projects yourself rather than having a professional do the work can be overwhelming at first. That's why we begin this book by discussing the system itself. It's here you'll realize that the whole is simply the sum of its parts and that with the proper groundwork you can attack this or that problem or tackle most plumbing improvements and come away successful.

Then, after a brief discussion of the tools of the plumbing trade you need to become familiar with, we move on to a section titled "Solving Plumbing Problems" where we discuss some common plumbing problems you are likely to have to deal with. Here you'll learn how to open clogged drains and to repair leaky and frozen pipes, faucets, toilets, and water heaters, as well as how to troubleshoot food waste disposers and to quiet noisy pipes.

The next major section of the book, "Making Plumbing Improvements," focuses on a good number of popular plumbing projects. In it you'll find step-by-step instructions and sketches showing how to install wall- and deck-mount faucets, fixture stop valves, lavatories and kitchen sinks, toilets, tubs, showers, and several more projects. And if you have some big plumbing plans, we've included information about extending existing supply and drain lines.

Next comes "Plumbing Basics and Procedures" where we tell how to choose the right pipe materials and fittings, how to measure pipes and fittings, and how to work with the various types of pipe. References to this basics section appear at appropriate junctures throughout the book.

Working to Code

Your responsibilities as an amateur plumber mirror those of a licensed tradesman—to provide for a supply of pure and wholesome water, and for safe passage of liquids, solid wastes, and gases to the outside. That means you must work to code, using only those techniques and materials regarded by the codes as acceptable.

The procedures in this book represent the editors' understanding of the 1979 Uniform Plumbing Code (UPC). (Canadian residents may obtain a copy of the Canadian Plumbing Code by writing the National Research Council of Canada, Ottawa, Ontario, Canada, K1A0R6.) Local codes and ordinances take precedence; so check with local officials to make sure you are complying with their guidelines. Check also to see whether your project requires a permit and any inspections.

GETTING TO KNOW YOUR SYSTEM

Because most of your home's plumbing components are hidden behind or under finish materials, about all you see are the fixtures and an occasional pipe disappearing into a wall or floor.

But these are only the tip of the plumbing system iceberg. The anatomy sketch at right depicts the elements common to all residential plumbing networks. Note that the *supply lines* and the *drain-waste-vent lines* act independently of each other.

Water enters your house by way of a sizable pipe that connects to a municipal water line or to a private well. If you have city water, it flows through a *meter* that monitors usage.

From there, it travels to the *water heater*. Water from a private system goes to a *pressure tank* before going to the heater.

From the water heater, a pair of water lines—one hot and one cold—branch out through the house to serve the various fixtures and water-using appliances. These supply lines are under pressure. Note the *stop valves* at the meter, the water heater, and the various fixtures and appliances. These enable you to shut down part or all of the system to make repairs or improvements. (If your lines aren't equipped with stop valves, see pages 48-49.)

The drain-waste-vent portion of your plumbing system depends on gravity to help rid the house of liquids and solid wastes. These lines also serve as a passageway to the outside for foul-smelling and potentially harmful gases. Note that all of the fixtures except the toilet empty into a *trap*. Water here

forms an airtight seal that prevents gases from backing up and leaking into the house.

From the trap, wastes travel through pipes sloped at no less than ¼ inch per foot to the larger *main drain*, then down and out of the house to a sanitary sewer, septic tank, or cesspool. Toilets, which are trapped internally, drain directly into the main drain.

One or more *cleanouts* in the main drain allow you to gain access for clearing clogged lines. Traps serve the same function for clearing fixture drains.

ANATOMY OF A PLUMBING SYSTEM

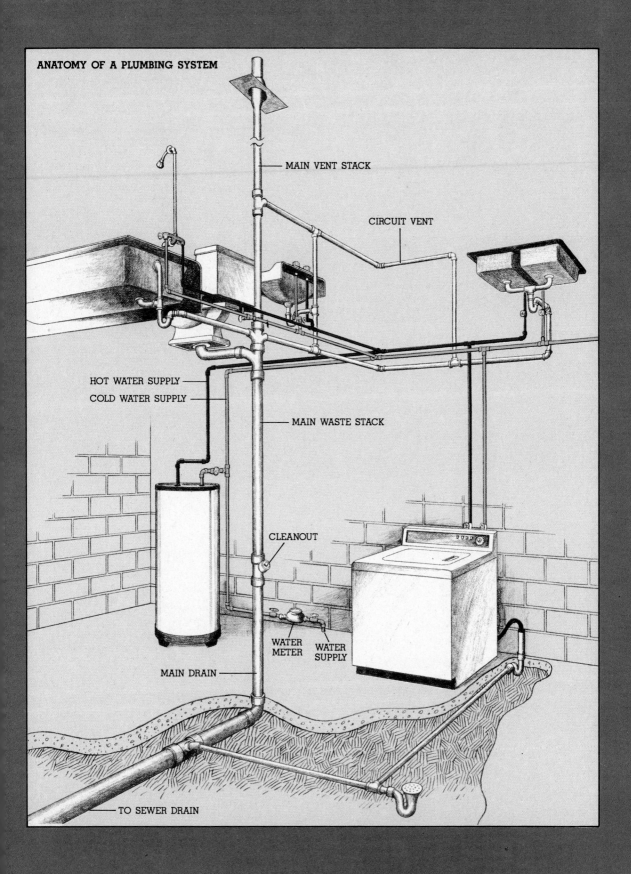

MAIN VENT STACK

CIRCUIT VENT

HOT WATER SUPPLY

COLD WATER SUPPLY

MAIN WASTE STACK

CLEANOUT

WATER METER

WATER SUPPLY

MAIN DRAIN

TO SEWER DRAIN

TOOLS FOR PLUMBING WORK

Ask anyone in the plumbing trade and he'll tell you that having the right tool for the job is a must. Without it, even simple tasks become difficult. The same applies to you as a do-it-yourself plumber. Fortunately, though, you probably have many of the tools shown here. Purchase the others if and when they're needed.

To clear drain lines—the most common of plumbing maladies—you'll need a *plunger* (or force cup), a *drain auger* for blockages that won't yield to plunging, and a pair of *rib-joint* or other type of *pliers* to remove the fixture's trap, if necessary. A *closet auger*, with its specially formed head, makes quick work of clearing blocked toilet traps.

For faucet and other repairs, make sure you have: *rib-joint pliers* or an *adjustable-end wrench* and a couple of *screwdrivers* (and possibly an *allen wrench*) to remove faucet handles, spouts, and seat washers; a *seat cutter* to renew pitted seats; and a *seat wrench* to remove hopelessly worn-out faucet seats. You'll find a *basin wrench* handy, too, as it allows easy access to otherwise hard-to-get-at nuts. In addition to these tools, have penetrating oil on hand to loosen stubborn nuts or screws. Also have an assortment of washers and O rings to replace defective ones.

Adding to, modifying, or repairing your home's plumbing lines calls for a different set of tools. *Calipers* help you determine both inside and outside pipe diameters. *Tubing cutters* make clean cuts in copper pipe and flexible copper tubing. For limited-space situations, you'll find a *mini-cutter* the ideal tool for cutting copper.

Bending flexible copper tubing to the desired shape is easy with a flexible *tubing bender*. And to prepare this material for a flare joint, you'll need a *flaring tool*.

The only tool required for soldering copper pipe is a *propane torch*. But to ready the pipe for soldering, you'll need emery cloth to remove oxidation, flux to allow for free flow of the solder and to aid the bonding process, and solder to seal the joint.

A *hacksaw* or a *close-quarters hacksaw* makes an easier chore of sawing through threaded or plastic pipe. And to disengage or join lengths of threaded pipe, get yourself a couple of *pipe wrenches*. Materials to have when working with threaded or plastic pipe are penetrating oil, joint compound or pipe tape, and solvent for joining plastic pipe.

And if you ever need to cut into cast-iron pipe, rent a *cutter* specially designed for this purpose. It is shown on page 72.

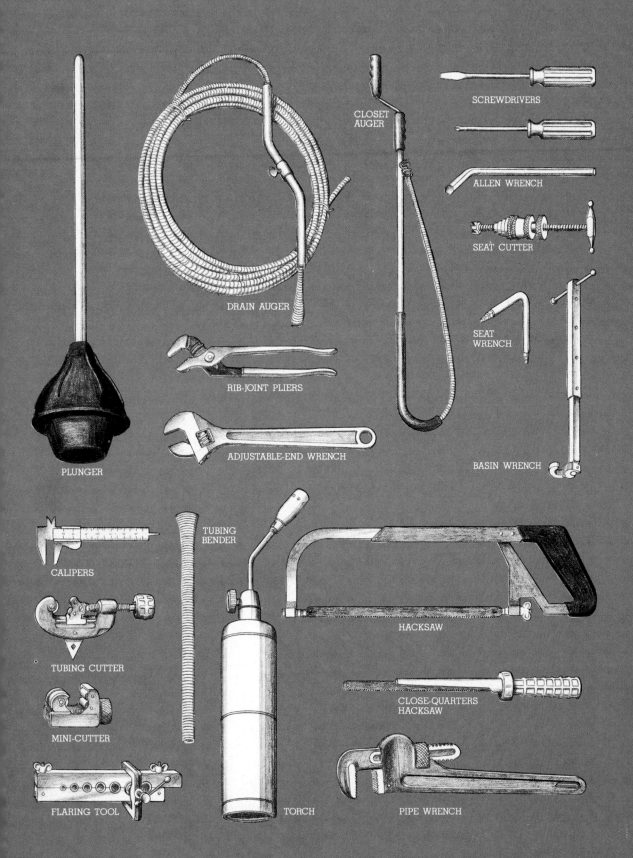

SCREWDRIVERS

CLOSET
AUGER

ALLEN WRENCH

SEAT CUTTER

DRAIN AUGER

SEAT
WRENCH

RIB-JOINT PLIERS

ADJUSTABLE-END WRENCH

PLUNGER

BASIN WRENCH

TUBING
BENDER

CALIPERS

HACKSAW

TUBING CUTTER

MINI-CUTTER

CLOSE-QUARTERS
HACKSAW

FLARING TOOL

TORCH

PIPE WRENCH

SOLVING PLUMBING PROBLEMS

Maybe you haven't been confronted yet by a stopped-up drain, a dripping faucet, or a gurgling toilet. But be assured: Sooner or later the law of plumbing averages will catch up with you.

Try to get someone to your home to fix one of these everyday nuisances, and you'll discover a second fact of contemporary life: The repair doesn't come cheap. When you have a plumbing contractor do the work for you, you're paying not only for his expertise, but also for a portion of the overhead involved in running his business. That's why it's not uncommon to pay $25 or more for even the simplest repair—one you may have been able to deal with yourself for a few cents.

With a few basic tools and elementary know-how, you often can handle many jobs yourself in fairly short order. For example, once you know of the existence of a *retrieving tool*, you can fish a ring or other item out of a drain as shown at left without the panic usually associated with these situations.

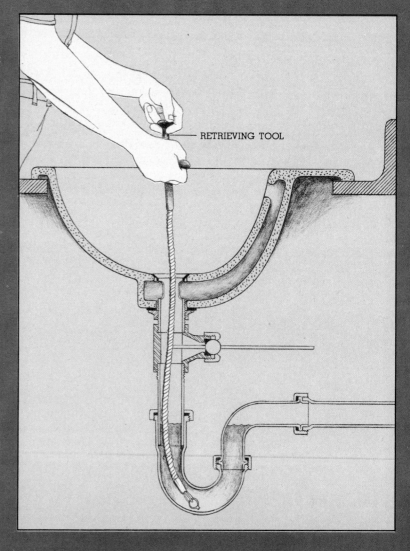

RETRIEVING TOOL

Opening Clogged Drains

When a fixture stops up, you'll naturally want to take immediate action. But before you rush at the problem with a plunger or auger, take a moment to analyze where the blockage seems to be.

Check the anatomy drawing on page 7, and note that your home has three types of drains.

Fixture drains have a trap and short sections of pipe on either side; *main drains* collect waste from all the fixture drains; and a *sewer drain* carries liquid and solid waste out of the house and to a community sewer, cesspool, or septic tank.

Nine times out of ten, the problem will be close to a fixture. To verify your suspicions, check other drains in your home. If more than one won't clear, something is stuck in a main drain. If no drains work, the problem is farther down the line, and you'll have to continue investigating.

Sinks and Lavatories

1 Clearing a sink or lavatory may involve nothing more than removing the strainer or stopper from the bowl's drain opening—a job that's generally as fast as it is easy. Bits of soap, hair, food matter, or other debris here may be the culprit.

Kitchen sink strainer baskets simply lift out. Some lavatory stoppers do, too. (See page 17 for a look at a typical lavatory.) Others require a slight turn before lifting. With a few, you must reach under the sink and remove a pivot rod.

2 A plunger uses water pressure to blast out obstructions. This means its rubber cup must seal tightly around the drain opening before you begin working the handle up and down. (Water in the bowl helps create this seal.) Stuff a rag in the overflow outlet of lavatories in order that the pressure can build and free the blocked passage.

3 If plunging doesn't work, fit an auger down the drain. Cranking its handle rotates a stiff spring that bores through a stubborn blockage. If this doesn't get results, dismantle the trap as shown on page 16, and auger the drainpipe that goes into the wall or floor.

Note: Chemical cleaners can sometimes speed up a slow-draining sink or lavatory, but don't dump them into one that's totally clogged. If they don't clear the drain, your problem is compounded by dangerously caustic water.

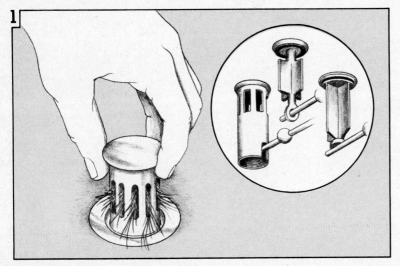

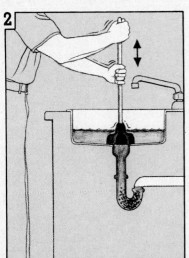

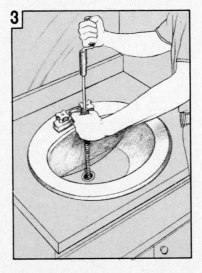

Opening Clogged Drains *(continued)*

Tubs and Showers

1 If a tub drain clogs, reach first for your plunger. If your tub has a *pop-up stopper*, you must remove it before plunging. Wiggling helps free the floppy linkage assembly.

Before you plunge, plug up the overflow, and allow an inch or so of water to accumulate in the tub (this helps seal the rubber cup around the tub outlet). As you work the plunger up and down, you will hear water surging back and forth in the drain.

2 If plunging doesn't do the trick, thread in an auger. If there's no stopper in evidence, you have a *trip-lever assembly*, like the one illustrated below, left. With this type, pry up or unscrew the *strainer* so you can insert the auger.

If you can get only a few inches of the auger into the drain and that doesn't clear it, then the problem is in the tub's trap directly below the overflow. To clear it, you'll have to follow a different route.

The best way to approach most tub traps is down through the *overflow tube.* This involves removing the pop-up or trip-lever assembly, which you do by unscrewing the overflow plate and pulling out the conglomeration of parts attached to it. (For more about pop-up and trip-lever assemblies, turn to page 17.)

Now feed the auger down through the overflow and into the trap. Cranking the auger all the way through the trap usually will clear the drain. If not, you'll have to remove the trap or a cleanout plug at its lowermost point, and auger toward the main drain. With a second-floor tub, this may involve making a hole in the ceiling below.

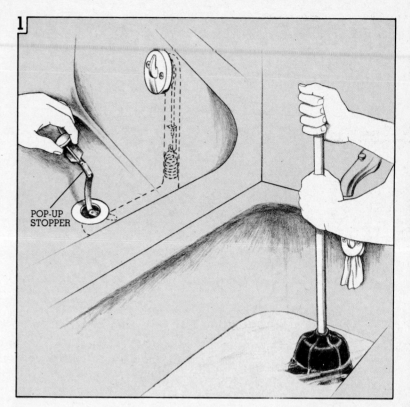

POP-UP STOPPER

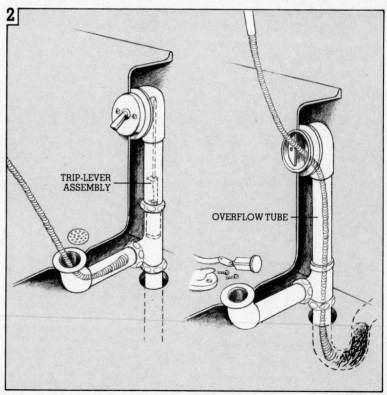

TRIP-LEVER ASSEMBLY

OVERFLOW TUBE

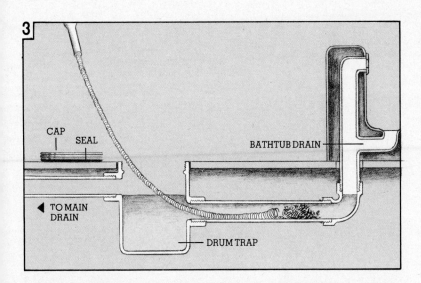

3 If there's a removable metal cap in the floor beside your tub, your tub is equipped with a *drum trap*. To free up one of these, begin by bailing out the tub; use rags or old towels to soak up any remaining water. Otherwise, the trap could flood over when you remove the cap.

Loosen it slowly, watching for water welling up around the threads. If this happens, mop up as you go. After you've removed the cap and its rubber seal, work the auger away from the tub, toward the main drain.

If, on the other hand, the trap is only partially full, as shown here, the obstruction is between the tub and trap, and you should auger toward the tub.

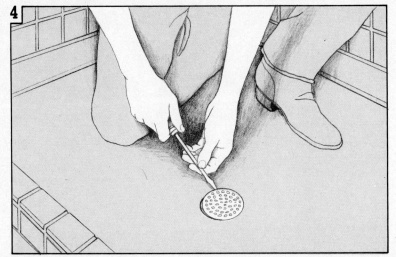

4 A clogged shower drain may respond to plunging. If not, remove its strainer, which may be secured to the drain opening by a screw in the center or snapped into place.

5 Now probe an auger down the drain and through its trap. If this doesn't work, you may be able to blast out the blockage with a hose. Pack rags around it, hold everything in place, then turn the water fully on and off a few times.

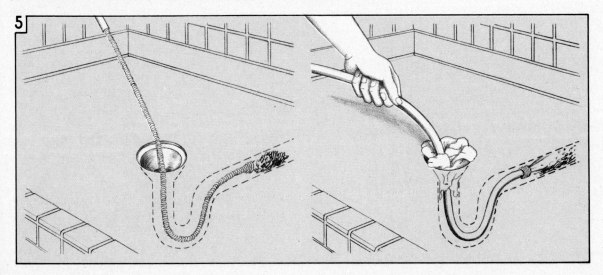

Opening Clogged Drains (continued)

Toilets

1 When a toilet clogs, don't flush it, or you'll also have a flood to deal with. Instead, use a bucket to carefully add or bail out water until the bowl is about half full. More than this could create a sloshy mess while you're plunging; too little, and the plunger won't make a tight seal around the bowl's outlet.

You can clear a toilet using an ordinary plunger, but the molded-cup type illustrated here generates stronger suction. Work up and down vigorously for about a dozen strokes, then yank away the plunger.

If the water disappears with a glug, you probably have succeeded. Check by pouring in more water. You may need to repeat the process several more times. If it doesn't get results, try augering, as shown in sketch 2.

Note: Never attempt to unclog a toilet with a chemical drain cleaner. Chances are, it won't do the job, and you'll be forced to plunge or auger through a strong lye solution that could burn your skin or eyes.

2 A *closet auger* makes short work of most toilet stoppages. This specialized instrument has a longer handle than the trap-and-drain version shown on the preceding pages.

To operate it, pull the spring all the way up into the handle, insert the bit into the bowl outlet, and begin cranking. If you encounter resistance, pull back slightly, wiggle the handle, and try again.

A closet auger will chew out just about anything but a solid object, such as a toy or makeup jar. If you can hear something other than the auger rattling around in there, you'll have to pull up the bowl, turn it over, and shake or poke out the item. See pages 54 and 55.

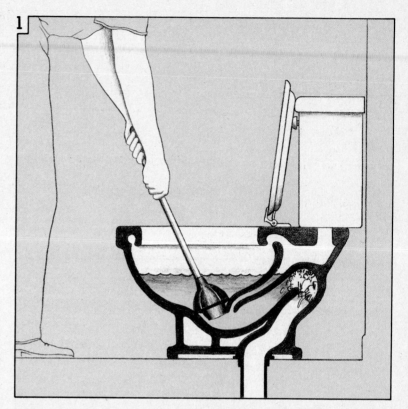

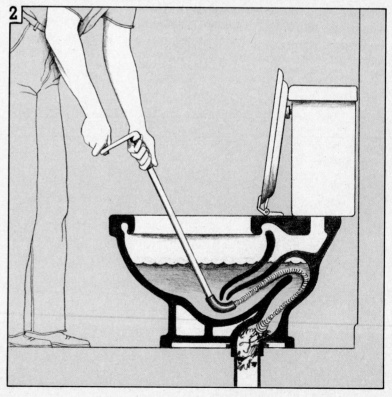

Main Drains and Sewer Lines

1 When one of these clogs, you may prefer to call in a plumber or drain-cleaning service. The work can be messy (you'll be dealing with raw sewage), and you may need a longer and stronger auger than the ones commonly used for fixture drains.

The key to getting an auger into a main drain is a Y-shaped fitting that is called a *cleanout*. You'll find one near the bottom of your home's soil stack, and there may be others higher up.

Begin by loosening the plug of that lowermost cleanout. If water oozes out, you can be sure the blockage is below somewhere. If not, try to find another cleanout above and work from there. Or climb up on the roof and auger down through the vent stack.

Before removing any cleanout, have buckets on hand to catch the waste water in the drain line.

Now thread an auger into the opening; work it back and forth a few times. Another often-successful way to clear main drains is to use a "blow bag" like the one illustrated.

2 If neither procedure works, you'll have to move downstream. Some houses have a house trap near where the drain lines leave the house. If yours does, open one of its plugs and thread an auger in. The blockage may be in the trap itself.

3 To clear a sewer line, try flushing it with a garden hose. Don't let the water run more than a minute or so, however; it could back up and cause drains to overflow. If this doesn't work, either call in a firm that specializes in opening clogged sewer lines, or rent a *power auger*. Operating one of these is a two-person job.

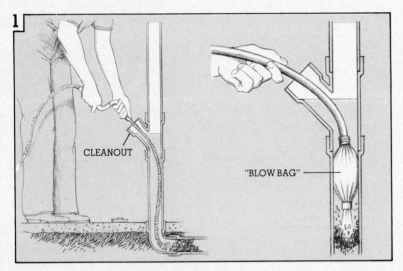

CLEANOUT

"BLOW BAG"

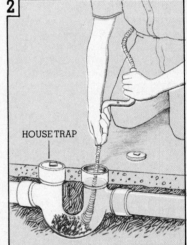

HOUSE TRAP

RAGS

POWER AUGER

Dismantling Fixture Traps

Sometimes the plunging and augering techniques shown on pages 11–13 fail to clear a drain. Or, sometimes you've dropped something down there and want it back. If either situation occurs, see whether the fixture's trap has a nutlike cleanout fitting at its bottommost point. Opening it lets you work an auger farther back toward the main drain or retrieve objects that have fallen in.

No cleanout? Don't be discouraged. It takes only a little more time and effort to remove the entire trap.

Trap configurations vary, but all include some combination of the slip-joint fittings shown below. These come apart easily, and let you reassemble components with a minimum of wrench work.

1 Before you begin, shut off water at the fixture stops (or the system shut-off) or remove faucet knobs so no one can inadvertently flood the scene down below. Position a bucket to catch water that will spill out when you remove the trap.

Now loosen any *slip nuts* securing the trap. Protect plating by wrapping tape around the jaws of your wrench or pliers. After a half-turn or so, you can unscrew the nuts by hand.

The exploded view here shows how *adjustable traps* dismantle. A *tailpiece* from the fixture slips into one end, an *elbow* connects the other end to a *drainpipe.* Most slip connections seal by compressing rubber washers; older ones may be packed instead with lamp wick, which looks like string, but makes a tighter seal. Other traps (not shown here) resemble J's and S's. They also come apart by loosening the slip nuts.

2 *Fixed traps* have slip fittings only at one end. To disconnect one of these, loosen both slip nuts, slide the tailpiece into the trap, then turn the trap loose from the drainpipe.

Before you reassemble a slip fitting, check its washer for wear or deterioration. Lamp wick always should be replaced; wrap a couple turns around before you begin threading on the nut.

Be careful, too, that you don't strip or overtighten a slip nut. Turn it as far as you can by hand, then use pliers or a wrench to go an additional quarter-revolution.

To test for leaks, completely fill the fixture, then open the drain and check all connections. Slightly tighten any that leak.

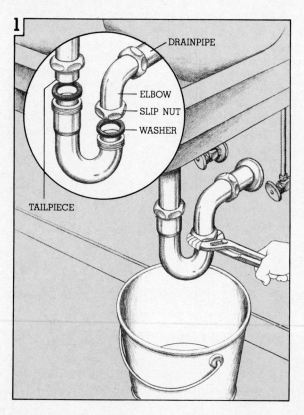

1
DRAINPIPE
ELBOW
SLIP NUT
WASHER
TAILPIECE

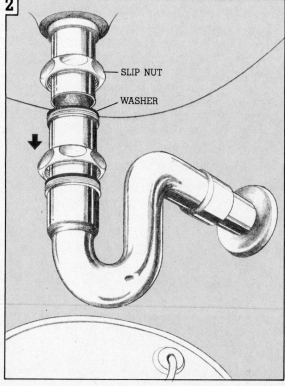

2
SLIP NUT
WASHER

Adjusting Drain Assemblies

When water in a lavatory or tub pulls a disappearing act or the fixture seems to take forever emptying itself, you can be fairly certain that a pop-up or trip-lever assembly isn't doing its job properly. With just a pair of pliers and a screwdriver, you can put a stop to either problem quickly.

1 If your tub or lavatory has a *pop-up* mechanism similar to those illustrated here, first pull out the stopper and thoroughly clean away any hair, soap, or other matter that may be keeping it from seating snugly. (See page 11 for help with removing lavatory stoppers.)

Next, check the *stopper seal*. If it's cracked or broken, pry off the rubber ring and install a new one. Look, too, for signs of wear or damage around the *flange* the stopper seats into.

Now replace the stopper and observe whether you can snug it down with the pop-up mechanism. If not, or if water is draining slowly, you need to make a simple adjustment or two.

For a lavatory, crouch under the basin and examine the position of the *pivot rod*. When the stopper is closed, this should slope slightly uphill from the *pivot* to the *clevis*. If it doesn't, loosen the *setscrew*, raise or lower the clevis on the *lift rod*, and retighten the screw.

Now the stopper may not operate as easily as it did before. This you can adjust by squeezing the *spring clip*, pulling the pivot rod out of the clevis, and reinserting it in the next higher or lower hole.

If water drips from the pivot, try tightening its *cap*. Or you may need to replace the *pivot seal* inside.

To adjust a tub pop-up, unscrew the *overflow plate*, withdraw the

entire assembly, and loosen the *adjusting nuts*. If the stopper doesn't seat tightly, move the *middle link* higher on the *striker rod;* if the tub is slow to drain, lower the link.

2 A *trip-lever* mechanism lifts and lowers a *seal plug* at the base of the overflow tube. When this drops into its *seat*, water from the tub drain can't get past. But because the plug is hollow, the overflow route is only slightly constricted.

Dismantle and adjust a trip-lever as you would a pop-up. Also check the *seal* on the bottom of the plug and replace it, if necessary.

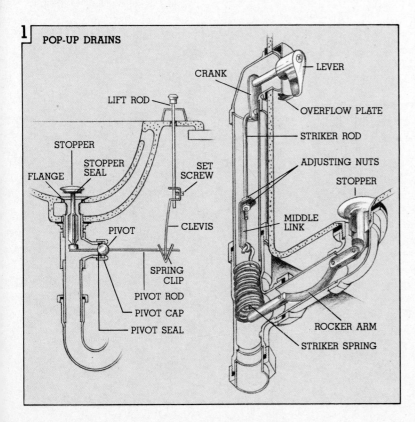

1 POP-UP DRAINS

CRANK — LEVER
LIFT ROD
OVERFLOW PLATE
STRIKER ROD
STOPPER
STOPPER SEAL
ADJUSTING NUTS
FLANGE
SET SCREW
STOPPER
CLEVIS
MIDDLE LINK
PIVOT
SPRING CLIP
PIVOT ROD
PIVOT CAP
PIVOT SEAL
ROCKER ARM
STRIKER SPRING

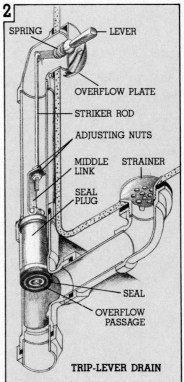

2 SPRING — LEVER
OVERFLOW PLATE
STRIKER ROD
ADJUSTING NUTS
MIDDLE LINK — STRAINER
SEAL PLUG
SEAL
OVERFLOW PASSAGE

TRIP-LEVER DRAIN

Remedies for Leaky and Frozen Pipes

If you've ever had a plumbing emergency at your house, you already know that water on the loose can wreak havoc. Even a tiny leak, left to drip day and night, will soon rot away everything in its vicinity. A pipe that freezes and bursts can cause a major flood when the thaw comes.

As soon as you spot a leak, shut off the water to take pressure off the line. Then locate exactly where the problem lies.

Water can run a considerable distance along the *outside* of a pipe, a floor joist, or the sub-floor, so it may take time and a strong light to find the problem's source.

Ultimately, any leaking pipe or fitting will need replacing. Pages 76–92 tell how. Meanwhile, unless you're dealing with a gusher or the problem is buried in a wall, floor, or ceiling, the temporary measures shown here will serve until you

can make a permanent repair.

Left unattended, any frozen pipe will turn into a leaking one, so you'll want to take immediate action when a freeze-up occurs. Again, these remedies will get you through a crisis but not necessarily prevent a recurrence. Pages 20 and 21 tell what to do about pipes that chronically freeze.

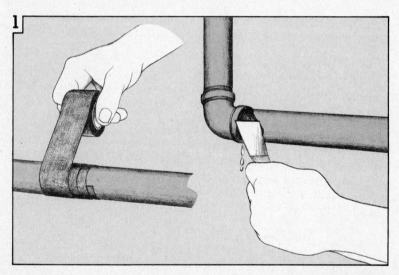

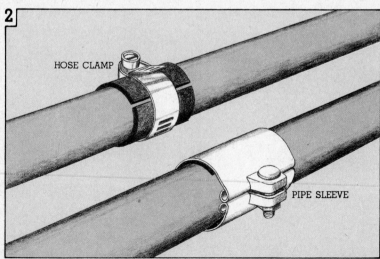

HOSE CLAMP

PIPE SLEEVE

1 For a pinhole leak, dry off the pipe and wrap it with several layers of plastic electrician's tape. Wind it about 6 inches in either direction of the hole.

At fittings, your best bet is to pack epoxy plumber's putty around the connection. This fast-setting compound makes a watertight patch.

2 An automotive *hose clamp* and a piece of rubber—both available at automobile service stations—also make an effective leak stopper. Just wrap the rubber around the pipe and tighten up the clamp.

The galvanized pipe commonly used in homes built a generation ago tends to rust from the inside out. Once a leak appears, you can expect others to follow. If the pipes at your house have begun to deteriorate, lay in a supply of *pipe sleeves* sized to fit your lines. These make semipermanent repairs that will last for several years.

If a leak seems to be more a drip than a squirt, and you can't find where it's coming from, the pipe simply may be sweating. Wrapping it with insulation, as shown on page 20, will eliminate condensation.

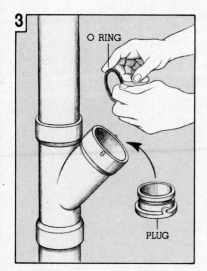

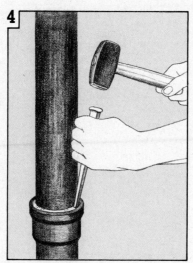

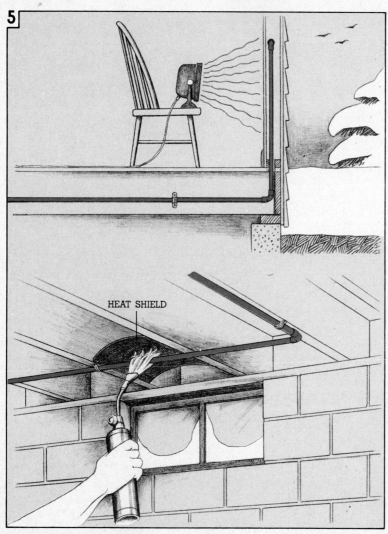

3 Drain-waste-vent (DWV) lines are less leak prone. Once in a while, however, a cleanout plug may begin to ooze water.

If this happens at your house, warn everyone in the household not to use any fixtures for a few minutes, then remove the plug and reseal it. For iron plugs, wrap the threads with pipe tape or coat them with joint compound. Plastic plugs twist free. Lubricate the O ring with petroleum jelly and replace the plug.

4 Leaks at the joints of cast-iron DWV pipes are easy to deal with. If yours is the hub-and-spigot type illustrated here, tamping down the soft lead it's been sealed with usually will eliminate the problem. Don't whack the pipe too hard, though; you could crack it.

Or perhaps your home's DWV lines are connected with no-hub clamping system depicted on pages 78 and 92. If so, simply tightening the clamp probably will stop the leak.

5 Frozen pipes obviously need to be warmed, and how you apply the heat depends to some extent upon where the pipe is. If it's concealed in a ceiling, wall, or floor, beam a heat lamp at the surface. Keep it 8 to 12 inches away so you don't risk starting a fire.

A propane torch offers the quickest (but riskiest) way to thaw exposed pipes. Never use one near gas lines, though, and use a heat shield to protect combustible materials. Ice tends to form along the entire length of a pipe, so put a spreader tip on the torch and move it back and forth. Don't let the pipe get too hot to touch; steam pressure could explode it.

If you don't have a torch or if the pipes are in tight quarters, wrap them with towels and pour hot water over the frozen section. Or heat the pipe with a hair dryer. Regardless of how you choose to thaw out a pipe, first open the faucet it supplies so steam can escape.

Preventing Pipe Freeze-ups

Icy-cold tap water may taste refreshing, but it's also a chilling omen that a pipe or pipes are in peril. Here are some steps you can take to bring them in from the cold.

1 Electric heat tape draws only modest amounts of current. You simply wrap it around the pipe and plug one end into an outlet. A thermostat turns the tape on and off as needed. Tape won't, of course, work during power outages, the times your home most needs protection against freezing temperatures.

2 Pipe jacketing comes in standard lengths you just cut with a knife and secure with plastic electrical tape. Ordinary insulation, cut in strips and bundled around pipes, works equally well. Be sure to insulate all joints and connections, too.

In an extremely cold wall or floor, you may be better off to pack the entire cavity with insulation. Also consider insulating long hot-water runs, especially any that pass through unheated spaces. You'll conserve water-heating energy.

3 As an emergency preventive, crack open the faucet you're concerned about and let water trickle through the line. If there's a cabinet underneath, open its doors and let room heat warm the pipes. Beaming a small lamp at the pipes also protects short runs through cold spaces during winter's worst.

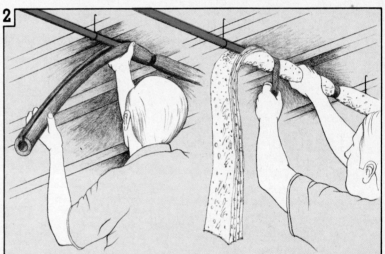

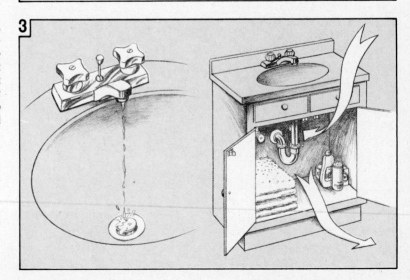

Winterizing Plumbing

Once upon a time, homeowners who were vacating a house for a month or two, or even the entire winter, simply turned down the thermostat and left enough heat to keep pipes from freezing. Today's high energy costs make that an expensive proposition.

Fortunately, you can shut down your entire plumbing system and let the furnace hibernate while you're gone.

1 Call the water department and ask them to turn off service at the valve outside your home. They may want to remove the meter, too. If not, close the valve on its supply side.

2 Now start at the top of your home's supply system and open every faucet. Shut off power to the water heater and drain it, too. At the bottom of the system, look for *stop-waste valves* near the water meter, and maybe elsewhere as well. Open the *drain cock* in each of these.

It's essential that you drain every bit of water from supply lines. If you find a low-lying pipe that doesn't have a faucet or drain cock, crack open a union, as shown on page 86. Siphon water from dish and clothes washers.

3 Empty fixture traps of water by pouring in automotive antifreeze mixed with water according to directions on the can. For a toilet, pour a gallon of antifreeze solution into the bowl to start the flushing action. Some of it will remain in the toilet's trap. Finally, if your house has a main house trap, fill it with full-strength antifreeze.

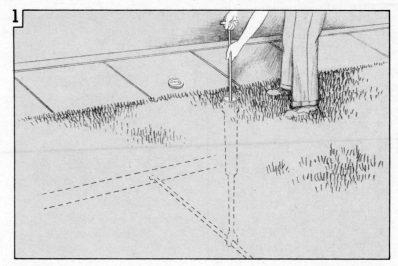

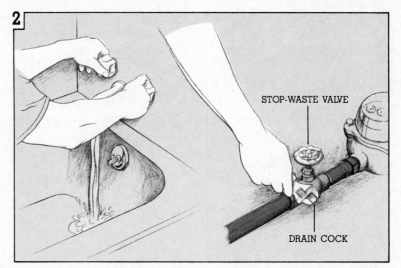

STOP-WASTE VALVE

DRAIN COCK

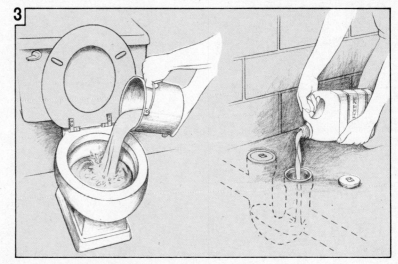

Repairing Leaky Faucets

A faucet's job is to deliver a stream of water on command, and rarely will yours fail to oblige whenever called upon. If trouble develops within, it almost invariably results in a drip, drip, drip from the spout, or an oozing from around the faucet body.

If either of these problems crops up around your house, you first must identify what type of faucet you're dealing with, then repair or replace the faulty part. Start by looking over the anatomy drawings shown here and on the following pages.

Stem Faucets

Stem faucets, such as the ones below, always have separate hot and cold controls. With many types, turning a handle twists a threaded *stem* up or down. In its off position, the stem compresses a rubberlike *washer* into a beveled *seat*, stopping the flow of water. As the washer wears, you have to apply more and more pressure to turn off the unit, and that's when dripping usually begins.

Newer versions, the so-called *washerless stem faucets*, replace the washer with a much more durable *diaphragm* or rubber *seal/spring assembly*. With this latter type, the stem rotates rather than raises and lowers to control water flow.

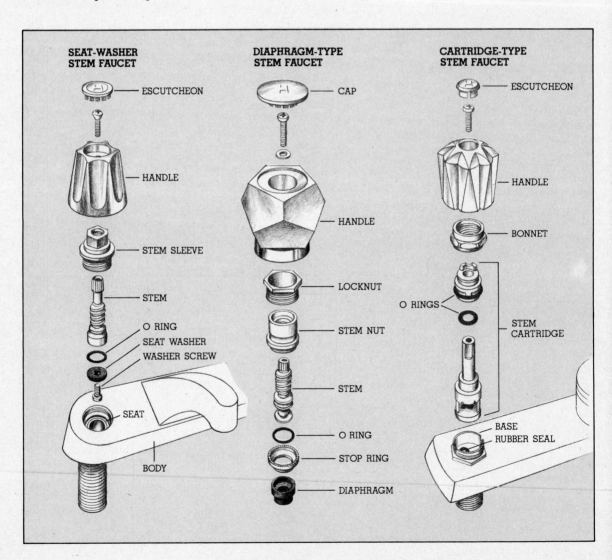

SEAT-WASHER STEM FAUCET
- ESCUTCHEON
- HANDLE
- STEM SLEEVE
- STEM
- O RING
- SEAT WASHER
- WASHER SCREW
- SEAT
- BODY

DIAPHRAGM-TYPE STEM FAUCET
- CAP
- HANDLE
- LOCKNUT
- STEM NUT
- STEM
- O RING
- STOP RING
- DIAPHRAGM

CARTRIDGE-TYPE STEM FAUCET
- ESCUTCHEON
- HANDLE
- BONNET
- O RINGS
- STEM CARTRIDGE
- BASE
- RUBBER SEAL

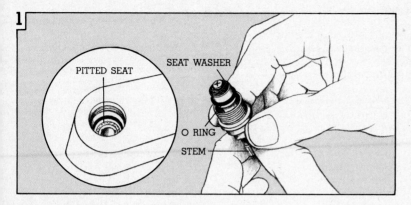

1

PITTED SEAT

SEAT WASHER

O RING

STEM

1 When the spout of a *threaded* stem faucet leaks, you can be sure that either the seat washer or the seat itself needs attention. Shut off the water supply to the faucet, then disassemble the faucet to the point where you can get a look at the washer at the base of the stem (see the anatomy drawing on the opposite page).

If the washer is cracked, grooved, or partially missing, back out the screw holding the washer to the stem and insert a new washer. Also check the O ring around the stem as well as the packing or packing washer. Replace these if needed.

Maybe the washer isn't the trouble maker. The seat at the base of the faucet body may be pitted or badly corroded.

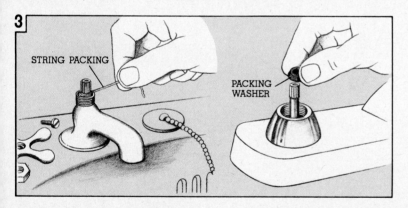

2

SEAT CUTTER

SEAT WRENCH

2 Depending on its condition, the seat may require either grinding or replacement. Special tools, a *seat cutter* and a *seat wrench*, perform these tasks. If you run up against a stubborn seat, squirt on some penetrating oil to free things up.

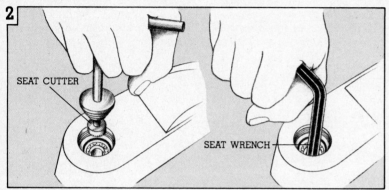

3

STRING PACKING

PACKING WASHER

3 You generally can trace leaks around faucet handles and from the base of the faucet to O rings or stem packing. Both wear out eventually and need replacement. Many older faucet stems came with packing that forms a tight seal under pressure. When replacing old packing, be sure to wrap the new packing clockwise around the stem. With newer faucets, a packing washer takes the place of the packing string.

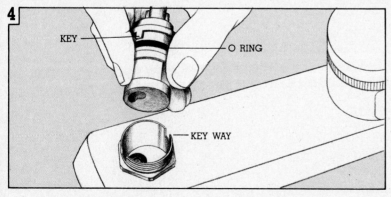

4

KEY

O RING

KEY WAY

4 If yours is the newer, cartridge-type stem, it's best to replace the seal and O rings whenever the faucet acts up. Remove the seal and spring with the end of a pencil. When reinserting the cartridge, be sure to align the *key* with the *key way*.

Repairing Leaky Faucets *(continued)*

Tipping-Valve Faucets

Its slim *control handle* makes a tipping-valve faucet easy to identify. This handle connects to a *control cam*, which when rotated activates the two *tipping-valve mechanisms* under the body cover. These mechanisms have several components: a *plug, gasket, stem, spring, screen,* and *seat.*

Raising the faucet handle forces the cam against the valve stems, lifting them off their seat. The farther back you throw the handle, the more water that enters the mixing chamber.

Note the handle's position in the sketch below. At left of center, it rotates the cam to tip the hot water stem, allowing only a stream of hot water to pass through. The cold water stem remains unaffected.

For the most part, tipping-valve faucet troubles—all easy to spot and correct—originate in three areas.

Leaks from the spout portend a breakdown of one or more of the valve mechanism components. Fix these by replacing the whole mechanism. If you notice leakage around the handle, the cam assembly O ring has given way. Sketch 3 on page 25 shows how to repair this. Finally, leaks around the spout mean a deteriorated spout O ring.

Though tipping-valve faucets no longer are made, you can still get the parts—sold in stores usually in kit form— necessary to remedy all of these problems. The repair kit won't include a screen; it's no longer considered needed.

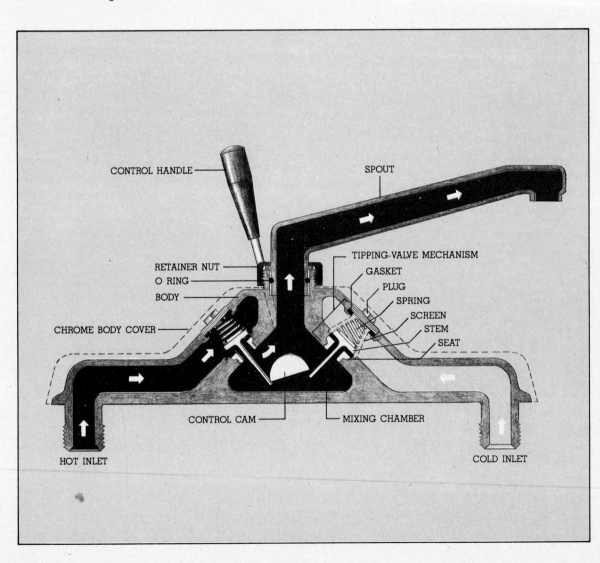

1 First things first. Shut off the water supply to the fixture and drain the water that remains at the faucet by raising the handle in its center position. To disassemble, loosen the retainer nut with a cloth-covered pair of pliers.

Next, grab hold of the spout and raise it out of the faucet body. If you spy a badly worn O ring, replace it with a new one.

2 To get at the cam assembly, you'll need to loosen the setscrew holding the handle to the cam. Then remove the rear closure concealing the cam.

Now remove the screws holding the cam in place, and pull it out of the faucet body. Set the cam aside. Lift off the body cover.

3 Using a wrench, remove the valve assembly plug, then the screen, gasket, spring, stem, and the seat. (You'll need an allen wrench or a seat wrench to remove the seat.) Replace the entire assembly with the repair parts. Do the same with the other valve.

4 After making all of the above repairs, you'll have one O ring—for the cam assembly—left. Simply remove the old O ring and replace it with the new. You would do well to lubricate the new one to make it easier to replace the cam in the faucet body.

If your faucet has a spray attachment that has been acting up, now's a good time to check on the condition of the diverter valve. It's under the spout in the faucet body and can be best removed using a screwdriver and pliers. Clean its openings with a toothbrush. If this doesn't help, buy a new one. When shopping for a diverter, be sure to take the old one with you so your local supplier can tell which type you have.

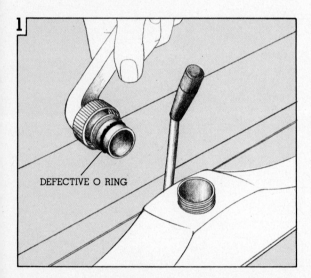

1
DEFECTIVE O RING

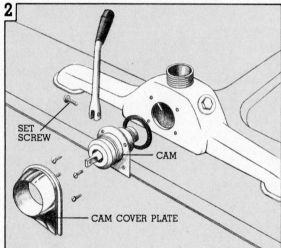

2
SET SCREW
CAM
CAM COVER PLATE

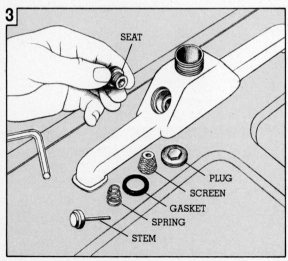

3
SEAT
PLUG
SCREEN
GASKET
SPRING
STEM

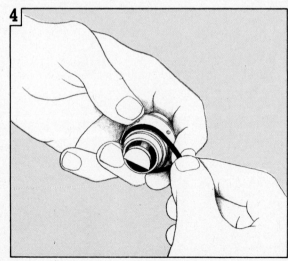

4

Repairing Leaky Faucets *(continued)*

Disc Faucets

As you can see by looking at the anatomy drawing below, disc faucets depend not on a washer and seat to shut off and control the flow of water, but rather on a *disc* arrangement.

Raising the faucet lever of one of these causes the upper portion of the disc assembly to slide across its lower half, allowing water to enter the *mixing chamber*. Naturally, the higher you raise the lever, the more water that enters. Conversely, lowering the lever closes off the *inlet ports*.

Moving the lever from side to side determines whether hot or cold water or a mixture of the two comes out of the spout.

The disc assembly itself, generally made of long-lasting ceramic material, rarely needs replacing. However, the inlet ports can become restricted by various mineral deposits. If this happens, simply disassemble the faucet as shown opposite and scrape away the minerals with a pocketknife.

If the faucet leaks around its base, one or more of the *inlet seals* probably needs replacing. It's a good idea to replace all of the seals, as the failure of one generally signals the impending demise of the rest.

Most plumbing supply outlets stock a supply of repair kits for this as well as other types of faucets. Before going to your supplier for a replacement kit, though, jot down the brand of faucet you have on a piece of paper (the name is on the faucet body). Or take the disc along if you already have removed it from the faucet body.

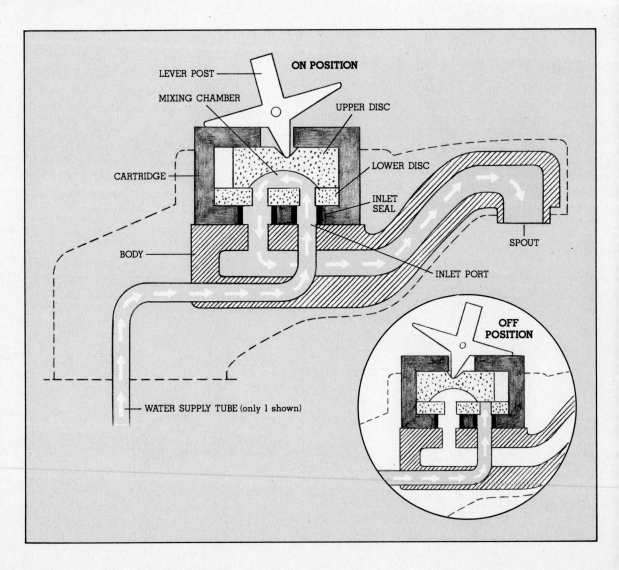

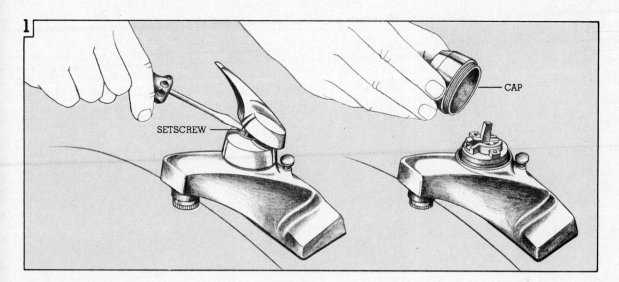

CAP

SETSCREW

1 To repair a leaky disc faucet, first shut off the water supply to the unit, then drain the lines by lifting the lever to its highest position. Look closely under the lever, and you'll see a setscrew that secures the lever to the lever post. Using an appropriately sized screwdriver, turn the setscrew counterclockwise until you can raise the lever off the post.

Next, lift off or unscrew the decorative cap concealing the cartridge. With this done, loosen the screws holding the cartridge to the faucet body, then lift out the cartridge.

2 On the underside of the cartridge you'll find a set of seals. To replace them, just pull out the old, worn-out ones and insert the new. While you're doing this, also check for sediment buildup around the inlet ports, and remove it to clear the restriction.

3 Reassemble the faucet, reversing the disassembly procedures. When inserting the cartridge, be sure to align its holes with those in the base of the faucet body.

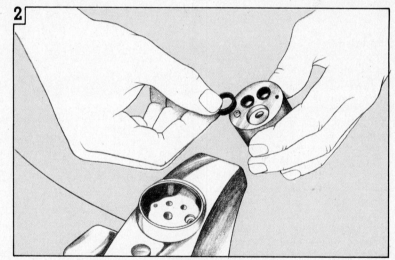

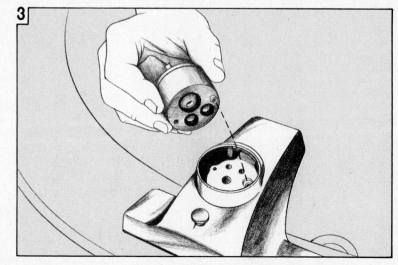

Repairing Leaky Faucets *(continued)*

Rotating-Ball Faucets

Inside every rotating-ball faucet, a slotted *ball* sits atop a pair of spring-loaded rubber *seals.* In the "off" position, this ball (held tight against the seals by the cap) effectively closes off the supply of water.

But look what happens when the faucet handle is raised (see detail drawings). The ball rotates in such a way that its openings begin to align with the supply line ports. When this happens, water can pass through the ball and on out the spout. Moving the handle to the left allows more hot water into the mixing chamber; to the right, more cold water.

Not surprisingly, usually after long use the seals and springs can give out. You'll find out how to replace these on page 29.

Realize, too, that these faucets can spring leaks from around the handle and, with swivel-spout models, from under the base of the spout. Handle leaks indicate either the *adjusting ring* has loosened a bit or the *seal* immediately above the ball has worn.

Under-spout leaks, on the other hand, result from O-ring failure. Inspect the rings encircling the *body* and, on units with diverter valves, the valve's O ring as well. Replace, if necessary, as shown opposite.

While you have the faucet apart, also check the ball for wear and corrosion. If it's faulty, simply replace it with a new one.

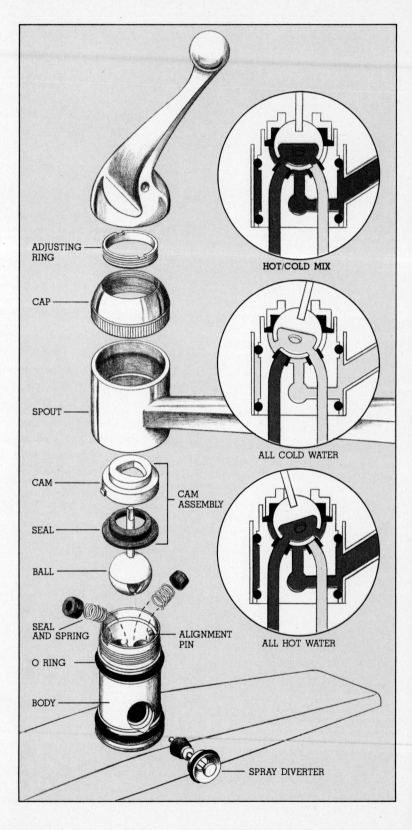

ADJUSTING RING

CAP

SPOUT

CAM

CAM ASSEMBLY

SEAL

BALL

SEAL AND SPRING

ALIGNMENT PIN

O RING

BODY

SPRAY DIVERTER

HOT/COLD MIX

ALL COLD WATER

ALL HOT WATER

1 To disassemble a rotating-ball faucet, first shut off the water supply, then drain the lines by lifting straight up on the handle. Using an allen wrench, loosen the setscrew holding the handle in place.

Next, loosen the adjusting ring (the wrench packed with repair kits is the correct tool), and unscrew the cap. You may need to apply pressure with cloth-covered adjustable pliers to budge the cap.

2 Lift out the cam assembly, ball, and, in the case of a swivel-spout faucet, the spout. The spout, being friction-fit around the body, may prove stubborn. So be prepared to apply some muscle at this point.

To remove worn seals and springs from the body, on the other hand, requires only minimal effort. Simply insert either end of a lead pencil into each seat, then withdraw the pencil. Check for restriction at the supply inlet ports, scrape away any buildup you find, then insert the new springs and seals.

3 If the faucet has a swivel-spout, pry the O-rings away from the body using an awl or other sharp-pointed tool. Roll the new ones down over the body until they rest in the appropriate grooves. Replace the diverter O ring in the same manner.

4 As you reassemble the faucet, be mindful that you must align the slot in the side of the ball with the pin inside the body. Note, too, that the key on the cam assembly fits into a corresponding notch in the body.

After hand-tightening the cap, tighten the adjusting ring for a good seal between the ball and cam. If there is a leak around the handle after restoring pressure to the lines, tighten the adjusting ring further.

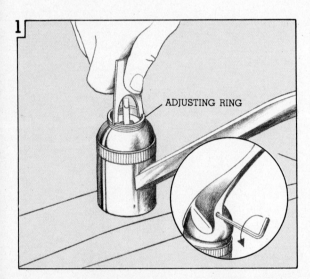

ADJUSTING RING

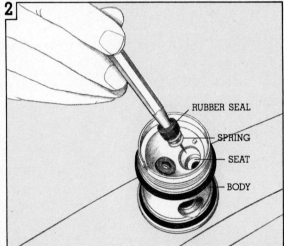

RUBBER SEAL
SPRING
SEAT
BODY

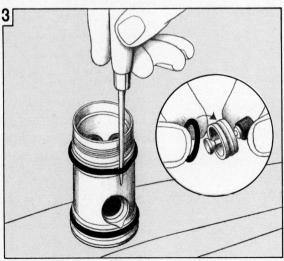

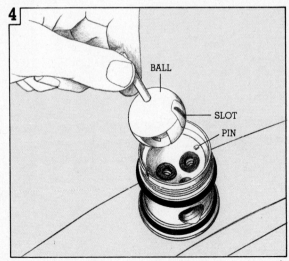

BALL
SLOT
PIN

Repairing Leaky Faucets *(continued)*

Sleeve-Cartridge Faucets

Most washerless faucets rely on a combination of seals and O rings to control and direct water. Not so with the sleeve-cartridge type. Instead, the cartridge itself is ringed by a series of strategically placed O rings.

Look at the anatomy drawing below, and you can see that the O rings fit snugly against the inside of the faucet body. This arrangement serves two purposes. The diagonally set O ring forms a seal between the hot and cold supply lines. The other O rings ensure against leaks from the spout, from under the handle, and on swivel-spout models from under the spout.

Note, too, that when the handle is raised, the stem raises also and the holes in it align with the openings in the cartridge. You control the temperature by rotating the handle either to the left (hot) or to the right (cold).

When this type of faucet acts up, you can replace either the O rings or the cartridge itself if it has corroded. And because of the faucet's simple design and few replaceable parts, repairing one generally doesn't take long. In fact, disassembling the faucet may account for the bulk of the work involved. Both of these procedures are covered on page 31.

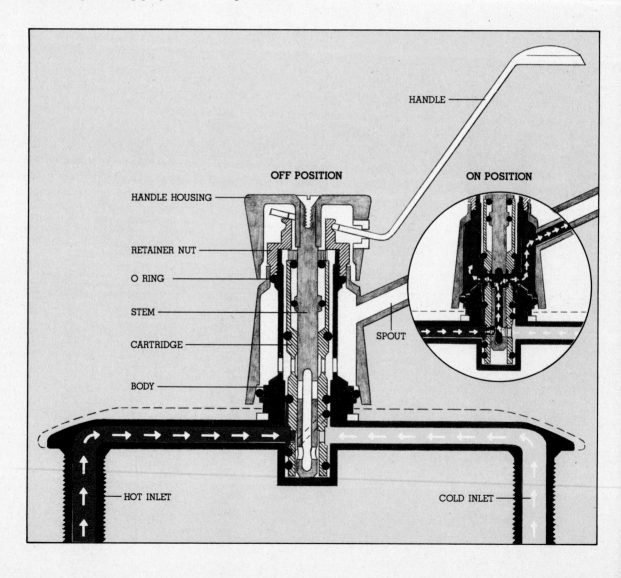

1 Sleeve-cartridge faucets vary somewhat in design from model to model, but all disassemble pretty much as follows. As always, you must shut off and drain the water lines first.

With this out of the way, pry off the decorative handle cover concealing the handle screw. Be careful you don't crack the cover in doing so; most are made of plastic.

Now remove the handle screw and lift off the handle assembly. On swivel-spout models, you'll encounter a *retainer nut.* Unscrew it, then lift off the spout.

Depending on the model you have, you may need to lift off a cylindrical sleeve to get at the cartridge. You should now be able to see the *retainer clip,* the device that holds the cartridge in place. Using long-nose pliers, remove the clip from its slot.

2 With pliers, lift the cartridge from the faucet body. Note the position of the *cartridge ears.* They face the front and back of the faucet, and it's important that when the cartridge is replaced they be in the exact same position.

3 Remove the O rings, install new ones, then reinsert the cartridge and retainer clip. If yours is a swivel-spout model, lubricate the O rings around the outside of the body, then force the spout down over the rings and into position.

4 Tighten down the retaining nut, using adhesive-bandaged or cloth-coated pliers to guard against marring the chrome spout. Finish the job by reinstalling the handle, restoring water pressure, and checking the faucet for leaks.

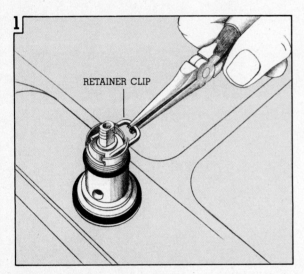

1

RETAINER CLIP

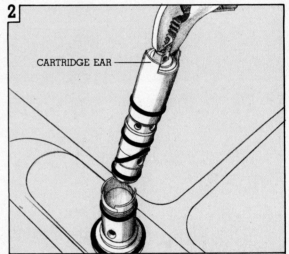

2

CARTRIDGE EAR

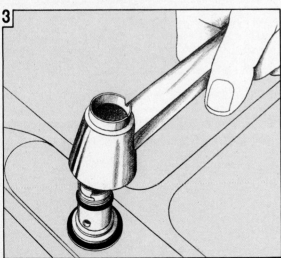

3

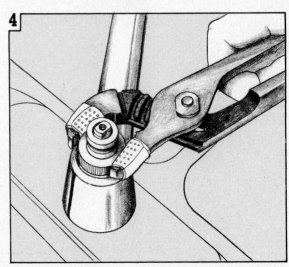

4

Repairing Leaky Faucets *(continued)*

Tub/Shower Faucets

Tub/shower faucet valve mechanisms—hidden behind the control handles—are so inconspicuous most people never give a thought to them. That's fine as long as everything works as expected.

When things go wrong, though, the faucet's sheltered position can cause you problems. A badly corroded faucet body or a leaky behind-the-wall supply connection in most instances means you'll have to perform wall surgery—a fairly involved operation—to get at and correct the problem. The text and sketches on the opposite page tell you how to handle these and other problems.

Though styled quite differently than faucets serving sinks and lavatories, tub/shower faucets function in much the same way (see the two examples below). Moving the handle of a stem faucet counterclockwise raises the stem out of its seat and allows water to pass through the faucet body. Some stem-type models have a third, diverter stem that directs water to the spout or shower head—whichever you wish. Others rely on a diverter spout to accomplish this. (See pages 34-35 for repairing diverters.)

With most single-control tub/shower faucets, activating the handle controls both the intensity and temperature of the water. Some of these have a diverter valve built into the faucet body; others depend on a diverter spout.

For information on how to rough-in a tub/shower faucet, see pages 56-57.

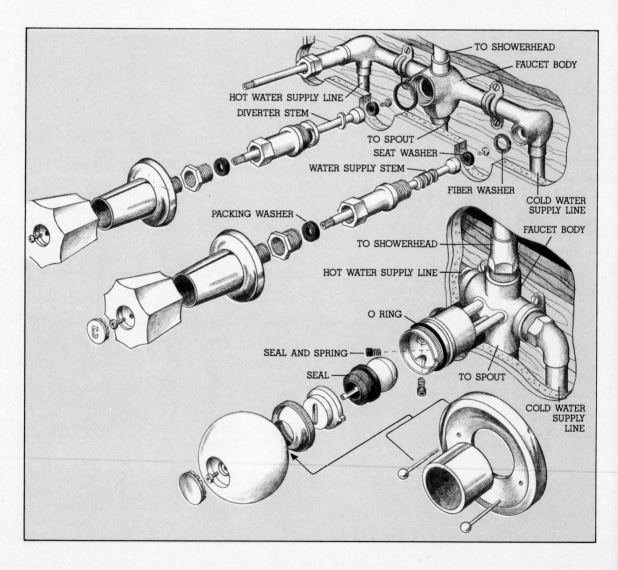

1 Spout leaks and leaks from around faucet handle(s) are just as common in tub/shower faucets as any other kind. When confronted with either of these problems, start by shutting off the water supply and draining the lines. Then, using the anatomy drawings on the opposite page as a general guide, disassemble the faucet to where you can remove the valve assembly.

Check the seat washer or the seals and springs at the base of single-control valves for wear. Note, too, the condition of the packing, packing washer, or O rings. Replace the necessary parts and reassemble the faucet.

2 Shower heads, too, sometimes need fixing. If water squirts out around the head, first try tightening the nut holding it to the arm. If it still leaks, then remove the head and check the washer for wear. Replace it if necessary.

If not enough water is coming out of the shower head, mineral deposits may be the villain. To investigate, disassemble the head, clean the orifices in it, then put everything back together again.

3 Over time, mineral deposits can choke off the flow of water through a faucet. Or maybe a connection between a supply line and the faucet begins to leak. Both situations require that you get at the faucet body. Before making any incisions, though, look around for an access panel. (Though not common in newer construction, older homes may have them.)

If necessary, cut into the wall, shut off the water supply, drain the lines, and cut the faucet body free with a hacksaw. If necessary, replace the supply lines. Hook up the new faucet and repair the wall surface.

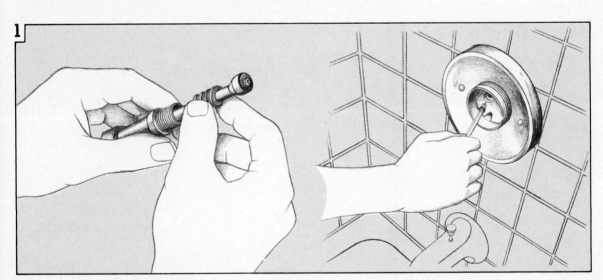

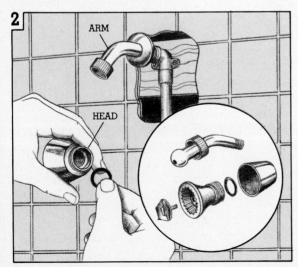

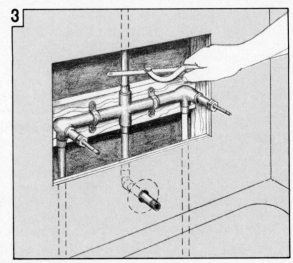

Repairing Sink Sprays, Diverters, Aerators

As mentioned earlier, some sink and lavatory faucets have sprays and diverter valves. And most all have an aerator at the tip of the spout. With sprays, troubles can develop in the connections, gaskets, or the nozzle. Most often, you can trace diverter maladies to worn washers or O rings. And about the only troubles that crop up with aerators are leaks caused by a worn gasket or a loose housing, and "low pressure" that results from mineral deposits clogging the screen. Fortunately, however, you can troubleshoot and correct all of these problems with a minimum of hassles.

1 Though diverters vary in shape from brand to brand, all operate in much the same way as the one shown here. When water isn't flowing toward the spray outlet, the valve remains open and allows water to proceed out the spout. But notice how it reacts when you press the nozzle's lever. It closes off the passage leading to the spout.

Nothing happens when you press the lever? Check to see whether the hose is kinked. A slow stream of water coupled with some water coming from the spout may signal a stuck valve or a worn washer or O ring. (See also step 2.) To check out the diverter, disassemble the faucet (see pages 24-31 for help with this). You'll find the diverter in the faucet body under the spout or in the spout itself. Replace the faulty parts, or the diverter itself, if necessary.

2 Minerals may be restricting the flow of water through the spray. Clean the spray disc with a straight pin as shown. Check other parts of the spray for wear and tighten all connections.

3 To check out a suspect aerator, disassemble it, then brush the screen clean if necessary.

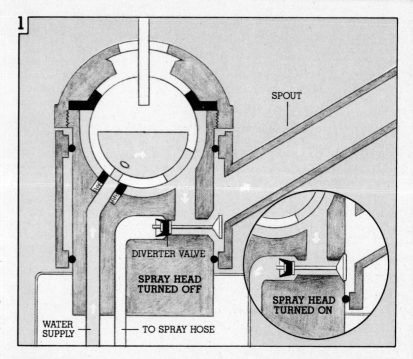

SPOUT

DIVERTER VALVE

SPRAY HEAD TURNED OFF

SPRAY HEAD TURNED ON

WATER SUPPLY

TO SPRAY HOSE

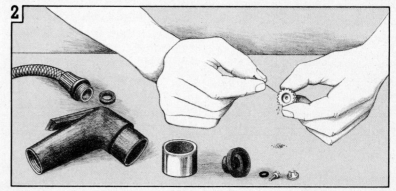

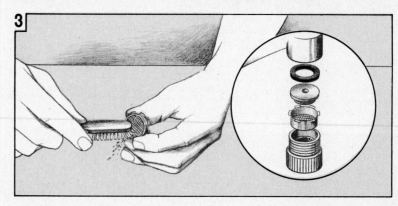

Repairing Tub/Shower Diverters

Tub/shower diverters fall into two general classifications. One group, typified by the stem type valve in the upper portion of the sketch, are housed in the faucet body and direct the flow of water from there. Tub diverter spouts, on the other hand, act independently of the faucet.

The sketch at right shows how each works. In the closed position, the diverter valve blocks off the water flow to the shower head. Opened fully, it diverts incoming water to the shower head. Here again, diverter mechanisms vary by manufacturer, but they all do the same thing.

With the tub diverter spout shown, lifting up on its knob while the water is running seals off the inlet to the spout and forces the water up to and out of the shower head. The water pressure will maintain the seal. But when the water is shut off the knob will drop back into its usual position.

When a tub diverter spout wears out, or if the lift rod attached to the knob breaks off from the plate it's attached to, you may as well replace the spout. To remove the defective one, insert a hammer handle or another suitable item into the spout and rotate it counterclockwise until it separates from the nipple it is attached to. Wrap pipe compound or tape around the nipple and install the new spout.

If a stem-type valve begins to leak or no longer will divert water properly, shut off the water supply to the faucet, drain the lines, and remove the nut holding the stem in place. Withdraw the stem, inspect the packing washer or O ring and the seat washer if your diverter has one, and replace any worn-out parts you find.

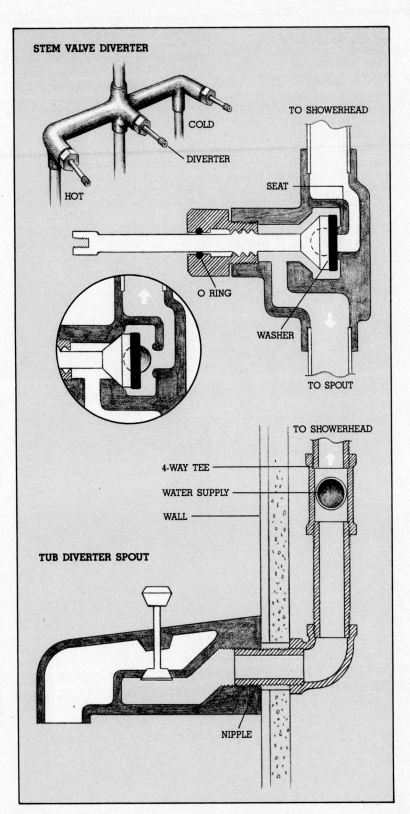

STEM VALVE DIVERTER

COLD

DIVERTER

HOT

TO SHOWERHEAD

SEAT

O RING

WASHER

TO SPOUT

TO SHOWERHEAD

4-WAY TEE

WATER SUPPLY

WALL

TUB DIVERTER SPOUT

NIPPLE

Repairing Toilets

Ever had to fiddle endlessly with the flush handle of a toilet to stop the water? Or fight with a tank ball that just refuses to rest squarely in its seat? Or peer helplessly into a toilet tank wondering what on earth is causing that incessant trickle of water? You're not alone!

Most people could care less about how toilets do what they do. But the sad fact is that some day you're going to have to make acquaintance with this necessary household unit. Let's take a behind-the-scenes look at a typical toilet and its rather simple workings.

When someone flips the *flush handle*, a chain-reaction of events occur. The *trip lever* lifts up the *tank ball* via a *lift wire/lift rod* arrangement. As the water rushes down through the *ball seat* and *flush passages* into the *bowl*, the reservoir of water and the waste in the bowl yield to gravity and pass through the toilet's *trap* out into a nearby drain line.

Inside the tank, the *float ball* rides the tide of the outrushing water until, at a predetermined level, the rod it attaches to trips the *flush valve.* (The tank ball settles back into its seat at this time, too.) This valve allows a new supply of water to enter the tank through a *fill tube* and the bowl through the *overflow tube.* When the float returns to its full position, the flush valve closes, completing the process.

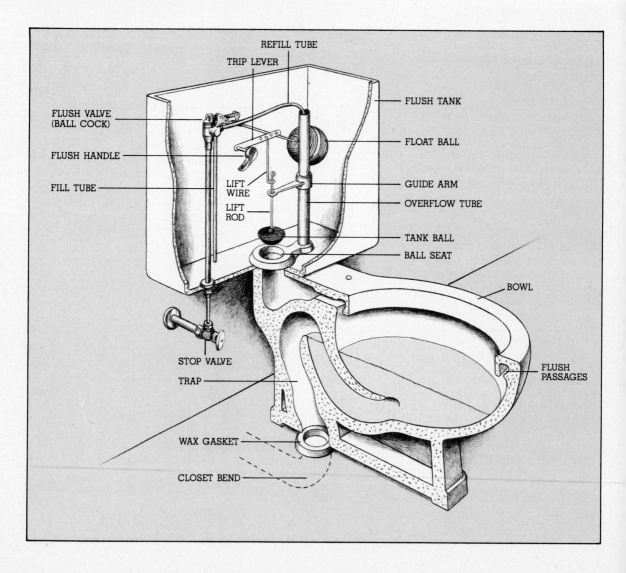

REFILL TUBE

TRIP LEVER

FLUSH VALVE (BALL COCK)

FLUSH HANDLE

FILL TUBE

LIFT WIRE

LIFT ROD

FLUSH TANK

FLOAT BALL

GUIDE ARM

OVERFLOW TUBE

TANK BALL

BALL SEAT

BOWL

STOP VALVE

TRAP

FLUSH PASSAGES

WAX GASKET

CLOSET BEND

Flush Tank Repairs

Since most of the mechanical action goes on inside the flush tank, it's not surprising that there, too, is where most problems develop. Here's a rundown of some common maladies and their solutions.

1 If you have difficulty getting the flush valve to close after a flush, the *float rod* may not be raising up high enough. Remove the tank cover, being careful not to chip it, and lift the rod with your hand. If the flush valve closes, bend the rod downward slightly. This simple procedure may solve your problem.

2 It's also possible the float ball has taken on some water. When this happens, the ball won't rise high enough to close the valve. To check out this possibility agitate the ball and listen for a swishing sound. To remove a faulty ball, rotate it counterclockwise until you disengage it from the float rod. Replace the ball with a new one.

3 If the float ball passes inspection, look next at the flush valve (ballcock) assembly. **Before attempting to remove the *float road mechanism*, though, shut off the water supply and flush the toilet.** Remove the thumbscrews holding the assembly in place, then lift it out and set it aside. (If yours is a diaphragm-type flush valve assembly, see the detail for help.)

4 Slip the blade of a screwdriver through the slot at the top of the plunger and lift it up out of the housing. Typically, you'll find a seat washer as well as one or more split washers. Remove and replace all of the washers, reassemble the flush valve assembly, and restore water pressure.

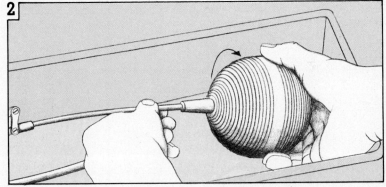

1 FLOAT BALL

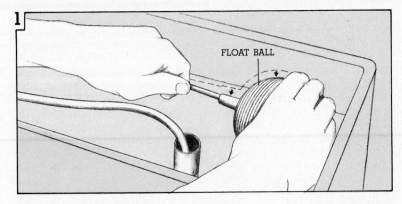

2

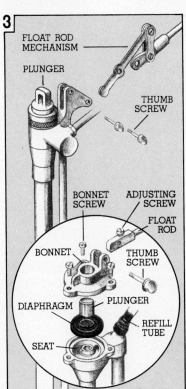

3 FLOAT ROD MECHANISM

PLUNGER

THUMB SCREW

BONNET SCREW

ADJUSTING SCREW

FLOAT ROD

BONNET

THUMB SCREW

PLUNGER

DIAPHRAGM

SEAT

REFILL TUBE

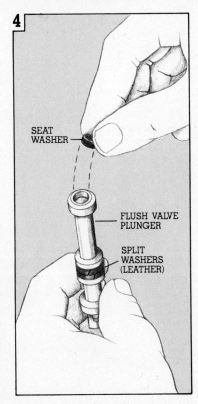

4

SEAT WASHER

FLUSH VALVE PLUNGER

SPLIT WASHERS (LEATHER)

Flush Tank Repairs *(continued)*

5 Of course, if for some reason water continues to leak out of the tank, the valve controlling incoming water may never get a chance to shut off. Or it may shut off only temporarily. If either situation fits your circumstance, and the float ball and flush valve check out properly, you have *tank ball* or *ball seat* problems.

Start by observing the tank ball as the tank empties. Does it settle squarely in its seat? If not, loosen the *guide arm* and rotate it for better alignment. Also check the *lift wire/lift rod* assembly and bend either part, if necessary, so the ball seats properly.

6 If the tank ball needs replacing, unscrew it from the lift rod and install a new one. Or, to eliminate any future problems with the lift wire, lift rod, or guide arm, replace the old ball with a *flapper* and chain. To do this, disengage the lift wire from the trip arm, then loosen and lift out the guide arm. Slip the flapper down over the overflow tube and fasten the chain to the trip lever.

7 A pitted or otherwise corroded ball seat also can prevent the tank from filling properly. To check out the seat, run a finger completely around it. If you detect a problem, scour the seat with a steel wool pad.

8 Newer flush valve assemblies, such as the one shown, simplify the flushing operation. And because they're corrosion-resistant plastic, they seldom act up. Sliding the float cup up or down on the rod controls the level of water in the tank.

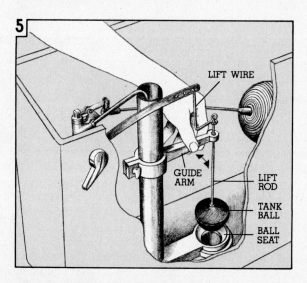

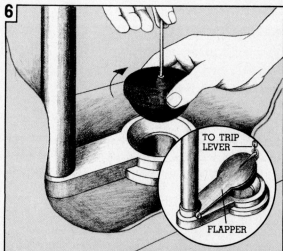

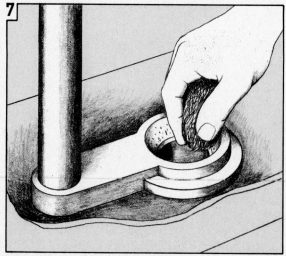

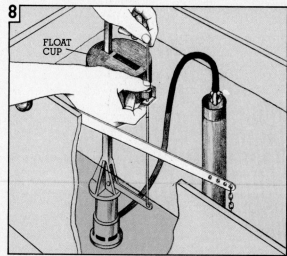

Stopping Exterior Leaks

1 As you can see, a *beveled gasket* at the base of the flushing valve shank and a *rubber washer* immediately beneath the tank (under pressure from a *locknut*) form a tight seal between the water inside the tank and the outside. Over time, however, either the locknut can work loose or the seals can give out.

If you notice a leak here, first tighten the locknut. If that doesn't work, shut off the water supply, flush the toilet, and sponge out the water that remains in the tank. Disconnect the water supply line, remove the locknut holding the flush valve as-sembly in place, and replace the old gasket and washer with new ones. Also, check the supply line washer for wear and replace it if necessary.

2 Extended use also can cause the *tank hold-down bolts* to loosen just enough to produce a leak. Remedying this is easy. Using a long-shanked screwdriver and a wrench, snug down the bolt as shown.

3 With some older-style toilets, the tank connects to the bowl via a fitting similar to the one shown. If leaks develop at either end of the fitting, tightening the nuts should dry things up in a hurry.

4 Leaks from around the base of the bowl indicate one of three things. The *bowl hold-down bolts* may need tightening, the *wax gasket* around the bowl inlet needs replacing, or the bowl is cracked. If a new gasket is in order, turn to pages 54 and 55 for more information.

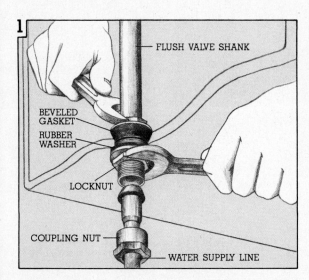

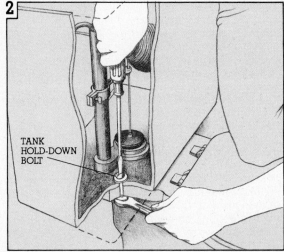

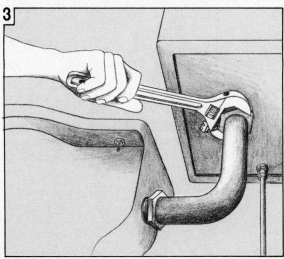

FLUSH VALVE SHANK
BEVELED GASKET
RUBBER WASHER
LOCKNUT
COUPLING NUT
WATER SUPPLY LINE

TANK HOLD-DOWN BOLT

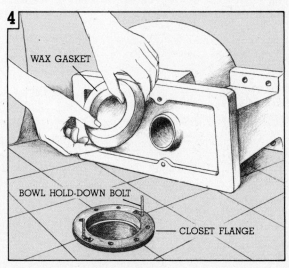

WAX GASKET
BOWL HOLD-DOWN BOLT
CLOSET FLANGE

Maintaining and Repairing Water Heaters

Study the anatomy drawings at right and you will discover that water heaters are little more than giant insulated water bottles. As hot water is used, cold water enters the unit via a *dip tube*. Naturally, this lowers the water temperature inside the tank. When this happens, a *thermostat* calls for heat. With gas and oil units, *burners* beneath the water tank kick in and continue heating the liquid until the desired temperature is reached. *Heating elements* perform the same function in electric water heaters.

Harmful by-products of combustion are ushered out of gas- and oil-fired units through a *flue* running up the middle of the tank. Electric heaters, since no combustion occurs, don't require venting to the outside.

When water heater troubles develop, it's usually due to sediment buildup or rust. You can do much to thwart the effects of both by opening the *drain valve* every few months and drawing off a few gallons of water. This purges rust and other gunk from the heater.

A regular maintenance checkup, about once a year, is your best insurance against most water heater maladies.

1 Either on top or high on the side of your water heater you'll find a *relief valve* to open if temperature or pressure inside the tank gets dangerously high. To test it, pull on its handle; if water rushes out of the pipe attached to it, all is well.

If nothing happens, close the stop valve in the cold water line, turn gas valve to pilot or shut off electrical power to the unit, and drain off enough water so the level in the tank is below the relief valve outlet. Now

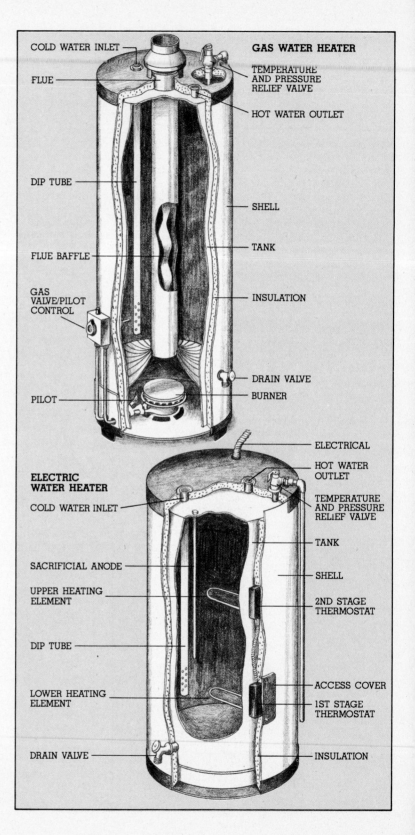

GAS WATER HEATER

COLD WATER INLET
FLUE
TEMPERATURE AND PRESSURE RELIEF VALVE
HOT WATER OUTLET
DIP TUBE
SHELL
TANK
FLUE BAFFLE
GAS VALVE/PILOT CONTROL
INSULATION
PILOT
DRAIN VALVE
BURNER

ELECTRIC WATER HEATER

ELECTRICAL
HOT WATER OUTLET
COLD WATER INLET
TEMPERATURE AND PRESSURE RELIEF VALVE
TANK
SACRIFICIAL ANODE
SHELL
UPPER HEATING ELEMENT
2ND STAGE THERMOSTAT
DIP TUBE
ACCESS COVER
LOWER HEATING ELEMENT
1ST STAGE THERMOSTAT
DRAIN VALVE
INSULATION

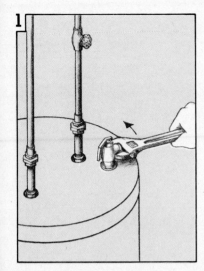

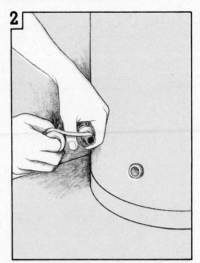

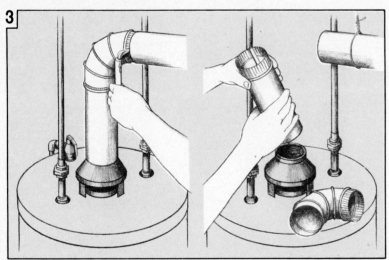

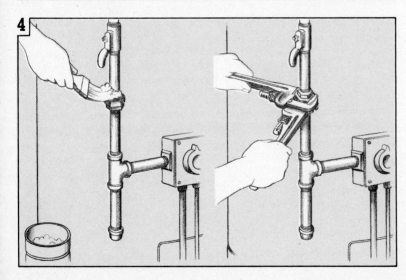

disengage the drain pipe from the valve, and the valve from the water heater.

Thread on a new valve, making sure to use pipe tape or pipe joint compound to seal the connection thoroughly. Reconnect the drain line, open the cold water stop valve, and you're all set.

2 Have a leaky water heater drain valve? To repair it, close the cold water stop valve, shut off the fuel supply, and completely drain the water heater. Screw out the faulty valve, then, after applying pipe tape or pipe joint compound to the connection's male threads, install a new assembly. Restore power to the unit, then open the stop valve and allow the tank to refill.

3 The condition of water heater flue pipes deserves careful monitoring. A rusted out or loose-fitting connection permits harmful vapors, including carbon monoxide, to enter the living area. If you suspect a leak in the water heater flue pipe, check out your suspicion by holding a candle as shown here. If the flame is drawn toward the pipe, you have a leak. This is one repair you shouldn't put off. Replace the defective section or sections immediately.

4 Water heater fuel line leaks are a serious matter, too. Generally, once a joint is sealed, you won't have any problem with it. But after making any repair to the fuel line, such as replacing a length of pipe, be sure to brush a soapy water solution on the joint. If you see new bubbles forming, there's a leak. Tighten the connection with a couple of pipe wrenches as shown in the illustration.

Troubleshooting Food Waste Disposers

A loud clanking noise, the strained buzz of an electric motor, or no action at all—each of these symptoms means you have waste disposer problems. Fortunately, though, the symptom often is worse than the illness. And if you know how to diagnose these strange goings-on, you should be able to get things going again without the expense of a plumber or a time-consuming trip to your plumbing parts supplier.

1 Disposers are pretty tough customers, but they're no match for flatware, bottle caps, and the like. If one of these undesirables falls accidentally into the grinding chamber, you'll hear the commotion right away. At best, the grinding blade will deform the item. Worse, a jam can result.

If your unit jams, shut off the power to it. Then remove the splashguard and survey the situation. Once you locate the obstruction, insert the end of a broom or mop handle into the grinding chamber and pry against the turntable until it rotates freely. With one brand of disposer, you insert an allen wrench into a hole in the bottom of the disposer, and work the tool back and forth. Remove the obstruction from the chamber. Impossible jams require professional attention.

2 If your disposer motor shuts off while in operation, its overload protector probably sensed overheating and broke electrical contact. To reactivate the motor, wait about five minutes for it to cool, then push the reset button (it's on the bottom of the disposer).

If the unit won't start, make sure the fuse or circuit breaker controlling the flow of power to the disposer is functioning. Verify, too, that the unit is plugged in or otherwise connected to the power source.

3 Since a disposer gobbles up huge amounts of food waste, it's only to be expected that occasionally the drain line may clog. If this happens, disassemble the trap (make sure you have a pan or bucket beneath it to catch the water that will spill).

If the trap itself is clear, thread a drain auger into the drainpipe.

Caution: Do not attempt to clear a blocked drain line with chemicals of any type because if the solution doesn't work, you'll have a line filled with caustic solution.

For information on how to install and operate a food waste disposer, see pages 58–59.

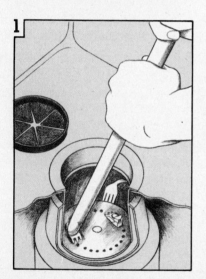

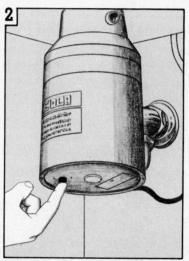

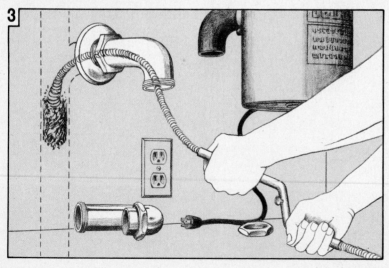

Quieting Noisy Pipes

Considering the conditions under which your home's water pipes operate, it's not surprising they make the noises they do. But that doesn't make their unexpected outbursts any less distracting. So if you've had it with that tick, tick, ticking, with the loud banging, and with all the other irritating clatter your pipes produce, read on.

Water hammer is perhaps the most common pipe noise of all. It results from a sudden stop in the flow of water, as would be the case when you turn off a fast-closing faucet.

You can generally trace *ticking* to a hot water pipe that was cool, then suddenly heated by circulating water.

Machine gun rattle, the annoying sound sometimes heard when you barely open a faucet, may indicate a seat washer is defective or loose. Air in the water lines also can be the culprit.

1 Generally, you can guess at the vicinity of a noisy pipe just by hearing it. So begin your sleuthing by going to the basement and checking to see whether one of the pipes has been knocking up against or rubbing a floor joist or subflooring. Once you've found the trouble spot, simply cushioning the pipe as shown here may be all you need to do. Short lengths of rubber pipe insulation are ideal for this.

2 The only sure way to deal effectively with water hammer is to install *water shock arrestors*, or *air chambers*, at strategic locations in your water lines. As you can see here, there are several ways to go, but the goal is always the same—namely, to provide a cushion of air for water to bang up against. Ideally, you should outfit each water supply line leading to each fixture with one of these devices. But to cut down on costs, start with those lines you know are causing problems. Installing a large air chamber between the water meter and the water heater makes good sense, too.

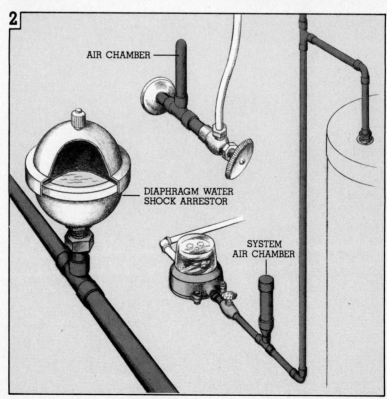

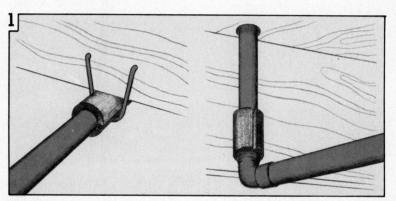

AIR CHAMBER

DIAPHRAGM WATER SHOCK ARRESTOR

SYSTEM AIR CHAMBER

MAKING PLUMBING IMPROVEMENTS

Too often people pick up their phone instead of their toolbox when it comes to installing a new water heater, sink, or even a faucet. Although usually the easy way out, there is another, less-expensive course of action. Why not tackle these and other plumbing improvements yourself? Sure, you'll take longer to make the hookups than a licensed plumber would, but think of the savings and the satisfaction you'll derive, not to mention the knowledge you'll gain about your home's plumbing system.

In the next 22 pages we have more than a dozen projects designed to improve your house's plumbing. You'll learn how to install faucets, sinks and lavatories, toilets, tubs, showers, water heaters, and much, much more. Naturally, some projects are more demanding than others, but you can accomplish every one of them without professional help.

Beginning on page 68, we even cover the specifics of how to extend existing plumbing lines to a new location. If this part of the project sounds as if it's too much to handle, you can always call in a licensed plumber to do the "rough-in," then take over from there yourself.

Installing Wall-Mount Faucets

Most faucets produced today are the *deck-mounted* type discussed on pages 46 and 47. But a great many *wall-mount* faucets still are in use, especially in older homes. So, many plumbing manufacturers carry a line of this type, too. Shown at right are two typical wall mounts: a newer type rotating-ball faucet and an older-style stem faucet.

Before going shopping for a new wall-mount faucet, measure the distance between the centers of the water supply pipes. Or better yet, take the old faucet with you. Doing either will enable your supplier to provide a compatible replacement.

1 Begin your installation by shutting off the water supply to the existing fixture and draining the lines. Then, loosen the nuts connecting the faucet body to the supply lines. On some newer models, you will have to remove the faucet body cover to get at the nuts. If you can't seem to budge

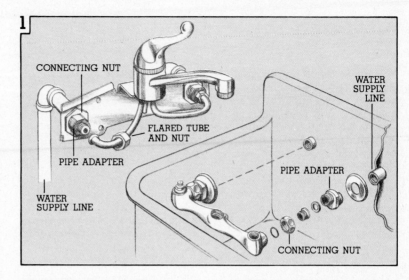

things loose, apply some penetrating oil, wait a few minutes, then try again. Set aside the old faucet body and unscrew all connecting hardware used to join the body to the supply lines.

Now follow the installation instructions that accompanied the new faucet. To ensure a leak-proof hookup, apply joint compound to or wrap pipe tape around all pipe threads.

When you've completed the installation, test for leaks by restoring water pressure. Tighten any loose connections.

2 A freezeproof wall hydrant, another wall-mount faucet, permits you to

run water to an exterior wall without fear of burst pipes. To install one, first tap into a nearby cold water line using a tee fitting. Then extend the run to about 12 inches from the exterior wall and thread, solvent-weld, or solder an adapter fitting to the pipe.

3 Bore a hole through the wall, slip the hydrant into place, and join it to the adapter. Outside, screw the escutcheon to the siding.

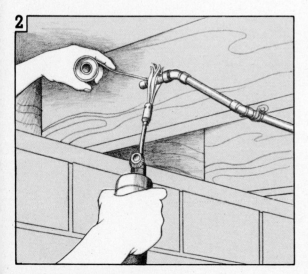

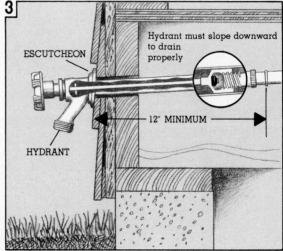

Installing Deck-Mount Faucets

The most difficult aspect of installing a new deck-mount faucet may be selecting the faucet itself. Your options are staggering, and as long as the faucet's inlet shanks align with the holes of the sink or lavatory you'll be securing it to, any style is fine. If you plan to install a sink or lavatory as well as a faucet, select the fixture first, then the faucet.

When replacing a worn-out or outdated stem faucet, ask your plumbing supplier for advice on the type faucet to purchase. Chances are he'll suggest one of the newer rotating-ball, disc, or washerless cartridge faucets because all are long-wearing and easy to repair.

For information on the correct way to install sinks and lavatories, refer to pages 50-53. And to find out more about the various types of faucets, see pages 22-31 for specifics.

1 Naturally, before installing a new faucet in an existing fixture, the old one must come out. To remove the old fixture, first shut off the water supply, then drain the lines. Work your way into the space below the sink or lavatory (this often can be quite a challenge); take a flashlight, a basin wrench, and an adjustable-end wrench with you.

Once you're in a fairly comfortable position, disengage the water supply inlet tubes from the faucet. Now, using the basin wrench, loosen and remove the locknuts holding the faucet to the deck. If your faucet is equipped with a spray attachment, remove the nuts securing the hose to the faucet body and the spray head to the deck.

After taking a well-deserved break, lift the old faucet up off the deck and set it aside.

Note: If your installation involves a sink or lavatory as well as a faucet, set the faucet and the drain assembly before positioning the fixture.

Begin installation of the new faucet by turning the faucet body upside down and slipping the *bottom plate* (smooth side down) and then the *rubber gasket* over the inlet shanks. Then set the faucet into the appropriate holes in the sink or lavatory deck.

2 Down under the fixture again, start a washer and locknut onto each inlet shank. Draw them up hand-tight, then tighten further with the basin wrench.

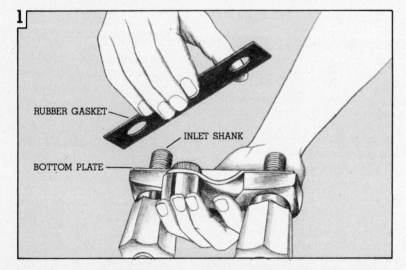

RUBBER GASKET

INLET SHANK

BOTTOM PLATE

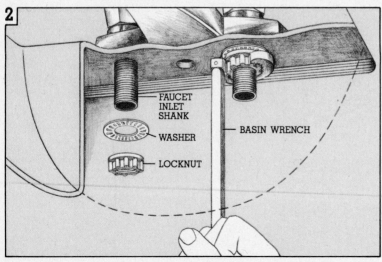

FAUCET INLET SHANK

WASHER

LOCKNUT

BASIN WRENCH

3 For faucets with spray attachments, secure the *hose guide* to the deck with the same washer/locknut arrangement used for the faucet itself. Then thread the spray hose down through the hole in the deck. Apply pipe joint compound to the threaded nipple at the end of the hose and secure it to the *spray outlet shank.*

4 To connect the water supply tubes to the faucet's inlet shanks, first fit a *compression nut, ring,* and *washer* onto each supply tube (this configuration varies from manufacturer to manufacturer). Then maneuver the tubes into the inlet shanks. Force the washer, ring, and nut up to each shank, then hand-tighten the nut. Further tighten with a wrench.

5 To complete the installation of a faucet in an existing lavatory, lower the *pop-up rod* down through the hole near the rear of the faucet spout, and through the holes at the upper end of the *linkage strap* attached to the *ball rod.* Tighten the thumbscrew to secure the rod.

With new lavatories, insert the *ball rod* into the opening in the *drain body* and secure it with the nut provided. Slip the rod through the linkage strap, then perform the procedure in the previous paragraph.

Open the water supply lines, partially fill the bowl with water, and check for leaks.

If the stopper fails to keep water in the bowl, loosen the thumbscrew and adjust the pop-up rod until the stopper provides a leakproof seal.

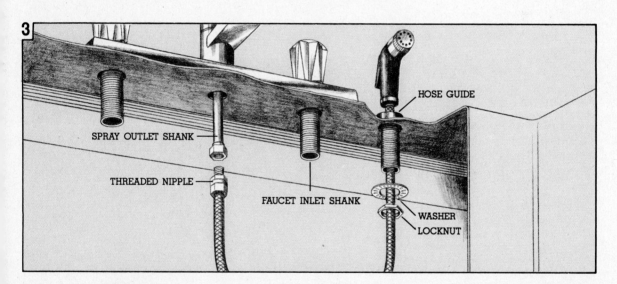

3
HOSE GUIDE
SPRAY OUTLET SHANK
THREADED NIPPLE
FAUCET INLET SHANK
WASHER
LOCKNUT

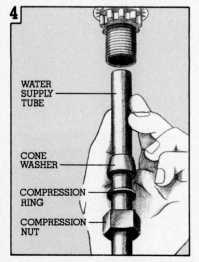

4
WATER SUPPLY TUBE
CONE WASHER
COMPRESSION RING
COMPRESSION NUT

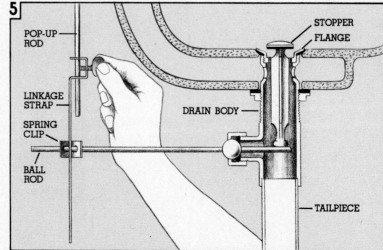

5
POP-UP ROD
LINKAGE STRAP
SPRING CLIP
BALL ROD
STOPPER
FLANGE
DRAIN BODY
TAILPIECE

Installing Stop Valves

Like most of the other elements that make up home plumbing systems, stop valves go largely unnoticed by most people. That is until a water line suddenly bursts. Or a faucet washer goes bad and causes a leak. Or it comes time to replace an old toilet.

At times like these, do-it-yourselfers gain a keen appreciation for these handy fittings.

If your home's fixtures are outfitted with stop valves, shutting off the water supply to them won't be a hassle, nor will it necessitate shutting down the rest of the system. (Just turn the valve handle clockwise till it's fully closed, open the faucet(s) to drain the lines, then go ahead with your job.) If not, consider adding them—a not-too-difficult task.

No matter what material your water supply lines are made of, there's a stop valve made to order, in sizes ranging from ¼ inch on up. With copper lines, use brass valves. Iron and plastic pipes take iron and plastic stops respectively. *Transition fittings* (see pages 78-79) even allow you to change materials (for example, from galvanized iron to plastic). Where the valve will be in view, choose one that has a chromed finish.

Plumbing outlets and building material home centers stock all the items you'll need. At some outlets, you can buy the components in kit form.

1 To determine your stop valve needs, simply take a quick look at your home's fixtures and other water-carrying equipment. Lavatories, sinks, tubs, showers, and clothes washers should have one for both hot and cold lines. Toilets and water heaters require only one, on the cold water line, and dishwashers need one on the hot line only. Check out the water meter, too. It should have a valve beyond it.

2 When you shop for a stop valve, you may well be asked whether you want a *gate* valve or a *globe* valve. In most residential installations the globe type is better for several reasons. It not only is reliable, but also can be repaired easily, if necessary, and can control the flow of water through the line.

Gate valves, on the other hand, are better suited to controlling main water lines and pump lines. As a result, you probably won't find many uses for them.

Note that globe valves operate in much the same way as stem faucets, whereas gate types control flow with a gate-like apparatus that moves up and down. Both types are available *straight* or *angled* to adapt to various situations.

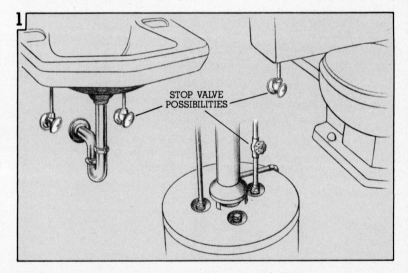

1 STOP VALVE POSSIBILITIES

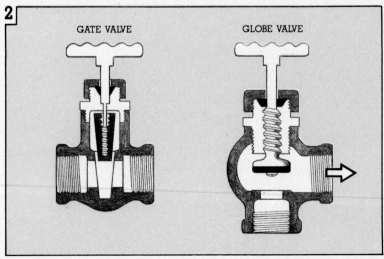

2 GATE VALVE   GLOBE VALVE

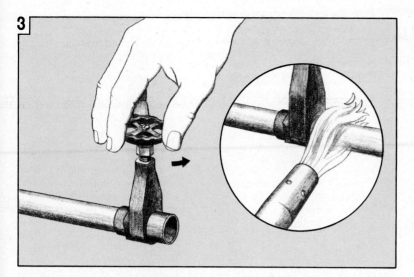

3 To install a stop valve, first shut off the water supply to the fixture and open the faucet to drain the line. Then cut into the line near the fixture. If you're dealing with copper pipe, use a tubing cutter; for iron or plastic pipe, you'll need a hacksaw.

Next, prepare the pipe to accept the fitting (see pages 82-90 for particulars). Before slipping a brass valve on a copper line, be sure to fully open the handle counterclockwise to prevent heat buildup and washer damage when heating the pipe. And when heating the connection for soldering, direct the flame at the valve, as shown, rather than at the pipe. Otherwise, the solder may seal the connection improperly.

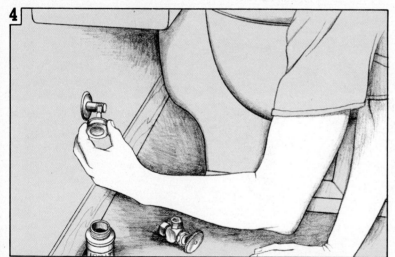

4 If you're adding a stop to a fixture supplied by threaded pipe, simply apply a liberal coat of joint compound to the pipe's threads, then screw the fitting on till it's finger-tight. Tighten the connection further with pliers and a pipe wrench. Protect chrome-finished fittings by wrapping the jaws of the pliers with adhesive bandages or scrap material of some sort.

With the connection complete, restore water pressure to the line and test for leaks.

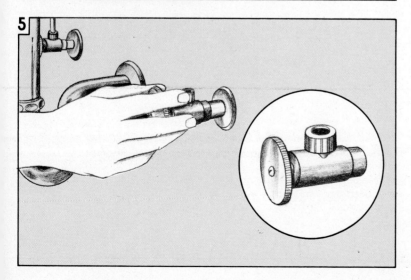

5 Installation of plastic stops differs only in that these you solvent-weld to the pipe. To do this, first dry-fit the stop and make alignment marks on the pipe and the valve. Remove any burrs that may be at the end of the pipe, and apply solvent to the inside of the fitting and the outside of the pipe. Working quickly, fit the stop on the pipe and give it a quarter-turn to spread the solvent. After making all hookups, turn on the water supply and test for leaks.

Installing Kitchen Sinks and Deck-Mount Lavatories

Ask any do-it-yourselfer who has ever completed a plumbing remodeling project and he'll tell you that the fun part comes when it's time to set the sink in place. It generally not only signals completion of the job, but also ranks as one of the easier plumbing tasks you'll undertake. See the material below and on the opposite page for all the particulars.

When shopping for a new deck-mount fixture, you won't find any shortage of products to choose from. What you will need to decide on is what material to choose—stainless or enameled steel, porcelain-covered cast iron, plastic, or vitreous china—and whether to buy a *rim-type* or *self-rimming* model (which has a rim-like flange aound the bowl).

Sometimes, you can save considerable money if you're willing to do a little comparison shopping. Just make sure that in doing so you compare apples with apples. Often, look-alike fixtures resemble each other in appearance only.

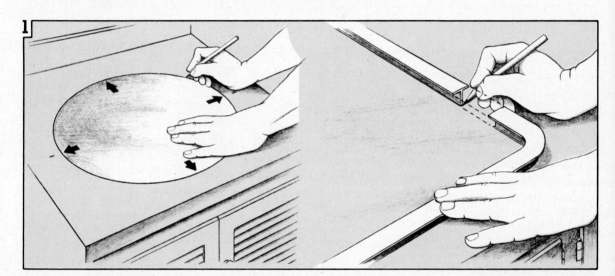

1 To replace an existing fixture, you must remove the old one. Begin by shutting off the water supply, then draining and disconnecting the water lines and the trap joining the sink or lavatory to the drainpipe. With self-rimming fixtures, you should be able to force the sink free by pushing up on it from below. Rim-type fixtures require that you first remove the lugs holding them in place. If you're removing a two-bowl cast-iron sink, you'll need help lifting it out of the opening.

To cut an opening in a counter top to accommodate a new sink or lavatory, first trace the outline of the opening. If yours is a self-rimming type fixture, use the template provid-

ed. With rim-type units, position the frame squarely, then trace around the outside edge of its leg as shown. Use a saber saw to make the cut in the counter top.

2 Before lowering the sink or lavatory into the opening, it's wise to make the faucet and drain assembly hookups. For help with mounting the faucet, see pages 46-47. With lavatories, the drain assembly consists of a *basin outlet flange*, a *drain body*, a *gasket*, a *locknut*, and a *tailpiece*.

To assemble the components, start by running a ring of plumber's putty around the basin outlet. Then insert the flange into the outlet. Complete the hookup as shown here.

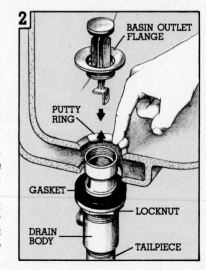

BASIN OUTLET FLANGE

PUTTY RING

GASKET

LOCKNUT

DRAIN BODY

TAILPIECE

3 Kitchen sinks, and certain other sinks, too, have basket strainers that tie the sink to the drain. To install one, lay a bead of plumber's putty around the sink outlet, then lower the basket (and a washer) into the outlet. With your free hand, slip the other washer and the locknut onto the strainer's shank. Tighten the locknut. Later, a tailpiece and trap will join the sink to the drain line.

4 To set a self-rimming fixture, apply a bead of silicone adhesive around the underside of the fixture's flange, about ¼ inch in from the edge. Then turn the fixture right side up and lower it into the opening, being care-ful to align it correctly. Press down on the fixture; some of the caulk will ooze out between it and the counter top. Smooth this excess with a slightly dampened finger.

5 To set a rim-type sink, fasten the rim to the sink, following the directions that accompany the rim. Now set the sink into the opening in the counter top and secure it with lugs positioned at 6- to 8-inch intervals around the sink (see the detail). Snug up the lugs with a screwdriver.

Complete the installation by hooking up the water lines as shown on pages 46-47, and the drain as shown on pages 16.

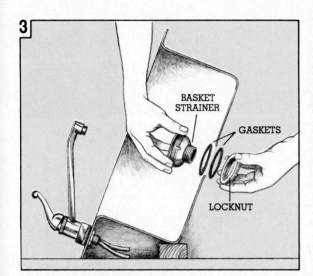

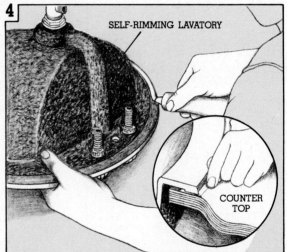

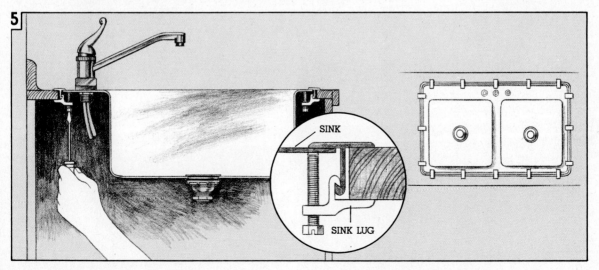

Installing Wall-Hung Lavatories

Like wall-mount faucets (see page 45), wall-hung lavatories are rapidly fading into the residential plumbing sunset. Most people prefer deck-mount fixtures like those on pages 50-51.

But plenty of wall-hung lavatories are still around. If you have one that needs replacing, or if you've had your eye on a pedestal-type lavatory at the local plumbing materials supplier, this and the following page will prove helpful to you.

As with their deck-mount cousins, wall-mount fixtures' prices tend to indicate the quality of the item.

1 When installing a new wall-mount, often your first job is to remove the old one. Start by shutting off the water supply; then drain and disconnect the water lines and the trap connecting the lavatory to the drainpipe. Pull straight up on the fixture and it should separate from the hanger bracket. If it doesn't, look underneath and make sure the lavatory isn't being held in place with bolts.

If yours is a new installation, running the water and drain lines comes first. For help with this part of the project, see pages 68-75. Once the lines are in, cut a 2×10 to span the distance between the studs that flank the pipes. Nail it between the studs. For greater strength, notch the studs to accept the 2×10 (the top of the 2×10 should be about 35 inches from the floor). Nail the 2×10 into place.

(With pedestal-type lavatories, which derive much of their support from the pedestal itself, you needn't bother with the 2×10. Special clips—supplied with the fixture—screwed firmly to the wall studs provide adequate support.)

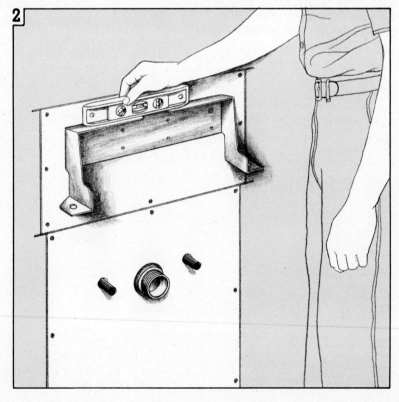

2 Now apply the finish wall material, then secure the hanger bracket to the 2×10 blocking, making sure that the bracket is level. (Refer to the instructions that accompany the fixture to find out the correct bracket height.) Use plenty of wood screws for this, as the bracket must withstand a considerable load.

3 Turn the lavatory on its side and make the faucet and drain body hookups. For help with this, see pages 46-47 and 50. Once this is out of the way, carefully lower the fixture onto the hanger bracket. The flange on the hanger bracket fits into a corresponding slot in the lavatory. With some models, toggle bolts help anchor the bracket to the fixture.

4 If the fixture you're installing came with support legs, fasten them to the lavatory, then adjust them to provide adequate support. To do this, twist the top portion of each leg. Use a torpedo level to check whether the fixture is level.

5 To complete the installation, first equip both water supply lines with a stop valve (if they don't already have them). Pages 48-49 show how to do this. Then connect the water supply inlets to the stop valves, and the trap to the drain body and the drainpipe. Restore water pressure, run water into the basin, and check for leaks. Tighten any loose connections.

    (Note: Bracket-supported wall-hung lavatories will stand up well to normal everyday use, but be sure to remind everyone in the family that exerting excessive downward pressure on them can spell trouble.)

3
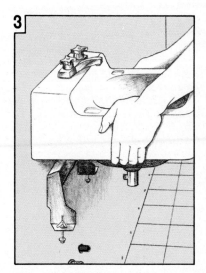

4
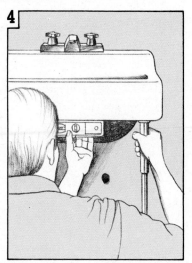

5
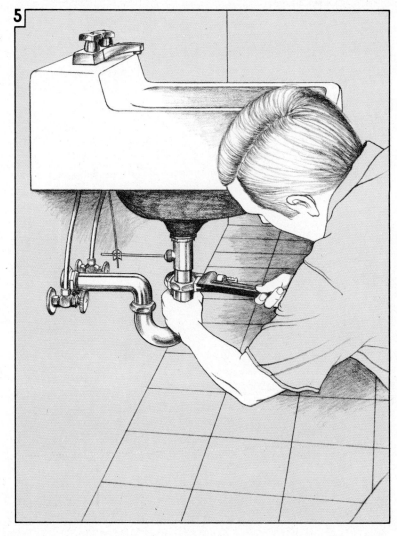

Installing a Toilet

Plumbers know it! Home builders and remodeling contractors do, too! But the average person has no idea how easy it is, under normal circumstances, to install a toilet. In fact, to replace an existing fixture or install one where the supply and drainage lines are in place, or "stubbed-in," should take you no more than a few hours to do the entire job.

Naturally, the job becomes more complex if you have to run water and drain lines to a new location. If that is your situation, refer to pages 68-75, which deal with extending existing plumbing lines. You may be best off to have a licensed plumber run the drain lines and possibly the water lines to the desired location for you.

The copy and sketches below and on the following page take you step-by-step from the installation of a closet flange onto a closet bend to the hookup of the water supply line.

When shopping for a toilet, you'll quickly notice the price differences between products. Most of this is attributable to quality variations. Though all toilets are molded of vitreous china clay, and baked in a high-temperature kiln, the similarity ends there.

Despite their attractive price, it's best to shy away from *washdown* toilets, as they are less efficient and noisier than the better-quality *reverse-trap* or *siphon-jet* models.

1 Unless you're replacing an existing fixture, start by installing a closet flange (be sure the bolt slots are parallel to the wall), then inserting the bowl hold-down bolts that come with the toilet. For plastic pipe, apply solvent to the outside of the *closet bend* and the inside of the *closet flange*. Fit the flange down onto the bend, then twist the flange a quarter-turn to spread the solvent. Further secure the flange by driving screws into the floor material below.

With lead pipe, cut the pipe down to floor level, fit a brass flange over the pipe, form the lead so it lies back against the flange (see the detail), clean the joint, apply flux, and solder the joint where the two materials meet (aim the torch at the flange rather than the lead). Again, fasten the flange to the floor with screws.

2 If you're replacing an old toilet with a new one, remove the old one using the techniques shown here, only in reverse order. Carefully remove the toilet from its container, then turn the bowl upside down on a cushioned surface. Run a bead of plumber's

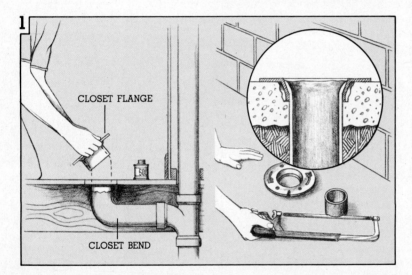

CLOSET FLANGE

CLOSET BEND

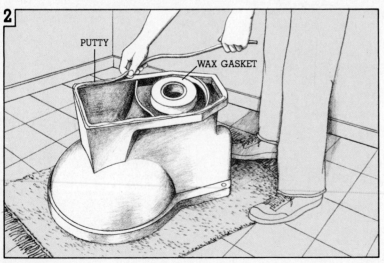

PUTTY

WAX GASKET

putty around the perimeter of the bowl's base, and fit a wax gasket (sold separately) over the outlet opening.

3 Return the bowl to its upright position, and gingerly set it in place atop the closet flange. Make sure the hold-down bolts align with the holes in the bowl's base. Slip a metal washer over each bolt, place on the nuts, and tighten snugly. Don't overtighten or you could crack the bowl and have to return to the store for another unit.

4 Now you're ready for the tank. First, lay the *spud washer*, beveled edge down, over the bowl inlet opening. This washer forms the seal between the tank and bowl.

5 Gently lower the tank onto the bowl, being sure to align the tank's holes with those toward the rear of the bowl. Now secure the tank to the bowl with hold-down bolts, washers, and nuts provided with the new toilet. Note that the rubber washer rests inside the tank under the bolt. Once firmly in place, this washer will prevent leaking.

6 Complete the installation by hooking up the water supply line as shown here, and then fastening the toilet seat (sold separately) to the bowl. Turn the stop valve to open, allow the tank to fill with water, then flush the toilet several times and check for leaks.

3

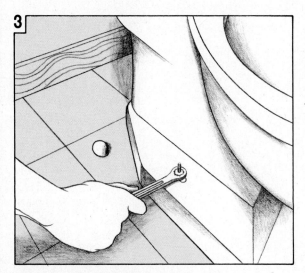

4

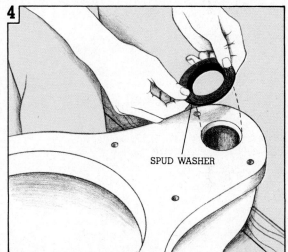

SPUD WASHER

5

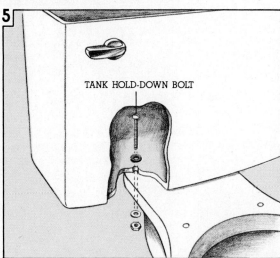

TANK HOLD-DOWN BOLT

6

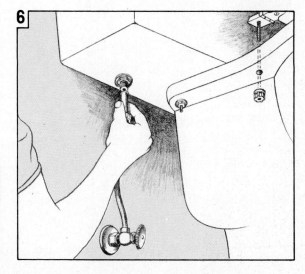

Installing Tubs and Showers

A successful tub or shower installation doesn't just happen. It requires careful planning, often some carpentry work, and a fair amount of time.

Start by thinking about where you would like to put the fixture. Then figure out how to get supply and drain lines there. For help with this, as well as for information on actually extending existing lines, turn to pages 68-75.

(Note: Sometimes, during house construction, plumbers "stub-in" plumbing lines for possible future use. You may be lucky enough to have the drain line you need poking up through your basement floor.)

Next, go to a local plumbing materials retail outlet and select a tub or shower and a compatible faucet and drain fitting. Standard tubs measure 5 feet long, but expect to find nonstandard sizes and shapes as well. Shower stalls also vary considerably in size and shape, though 32- and 36-inch units are typical.

You have a choice of materials, too. You'll encounter enameled cast iron, enameled steel, and fiber glass-reinforced tubs, and fiber glass and light metal shower stalls. And if you're planning a tile-lined or other hard-surface material-lined shower, you can purchase just the base in a variety of attractive materials.

The sketches and directions below and on the opposite page show and describe three typical installations.

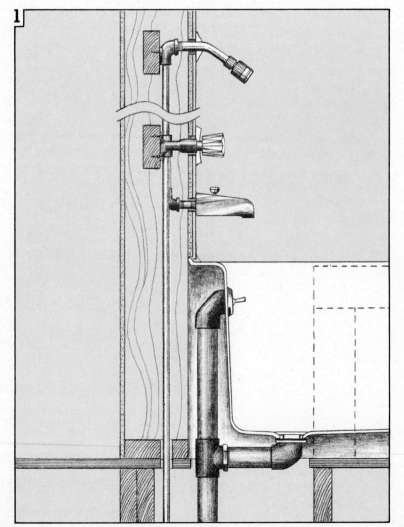

1 Begin tub installation by building an enclosure to house the unit. Then run supply and drain lines to the tub. Note that here the water supply lines come up through the soleplate. You'll need to cut a large enough opening in the subfloor to make the drain line connections.

Mount the faucet body according to the instructions that came with the faucet, then connect the supply lines to the faucet. If yours is a tub/shower installation, run the water supply up to the shower head.

Now slide the tub into place. The back of a cast-iron or an enameled steel tub should rest on a 2×4 support frame. The weight of a cast-iron tub will keep it in place. Anchor enameled steel and plastic tubs to the enclosure by nailing through their flanges into studs.

Make the drain assembly connections, test for leaks in the supply and drain connections, then surface the walls and ceiling of the enclosure with appropriate material. If you use drywall, it should be the moisture-resistant type. Apply silicone caulk around all openings and between the wall and tub to seal against moisture. Complete the project by installing the tub spout, faucet handles and trim, and the shower arm and head.

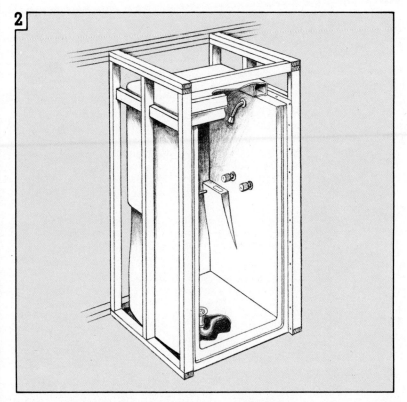

2 To build in a fiber glass-reinforced shower, first construct the frame that will surround it. Make sure you square the walls and position the frame so that when you slide the shower into the enclosure the drain outlet will be directly above the drainpipe. The instructions accompanying your unit should indicate the correct dimensions. Be sure also to back the nailing flanges at the front and top edges of the shower with framing so you can anchor the unit securely in place.

Then after fastening the drain fitting to the shower outlet, slide the unit into its frame. The fitting should form a tight seal between shower outlet and the drain. (Note: To provide firm support under the shower base, you may want to spread a sizable pile of concrete around the drain line first, then set the shower down into it.) Check for plumb, then secure the shower to the framing.

Drill holes (from the finished surface) for the faucet and shower arm, and mount the faucet. Again refer to the installation instructions for exact locations and how to. Now run the water supply lines to the faucet and on to the shower head.

Check for leaks. If all is well, finish the walls; caulk around all openings; install the faucet handle, shower arm, and shower head; and install a shower door.

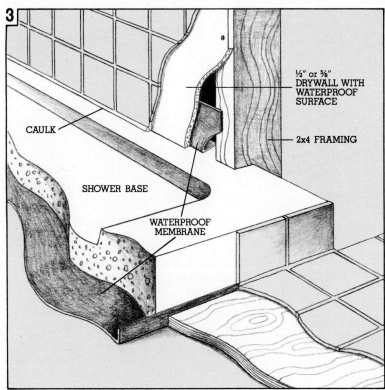

CAULK

SHOWER BASE

WATERPROOF MEMBRANE

½" or ⅝" DRYWALL WITH WATERPROOF SURFACE

2x4 FRAMING

3 This cutaway drawing shows what's involved with a tile-walled shower. Note that the shower base rests on a waterproof membrane. This same material protects the enclosure's notched framing members. Except for the way you frame and surface the enclosure, the installation closely resembles that of the plastic shower described in caption 2.

Installing a Food Waste Disposer

Of all the labor-saving home appliances, one of the hardest working is a food waste disposer, the little giant that takes in and devours many times its weight in food waste each week.

Installing one takes several hours, but the work involved isn't all that difficult. Here and on the next page we'll walk through the whole process.

To ensure satisfactory operation of the unit, you'll want to keep the following in mind. Whenever you operate the disposer, first turn on the *cold* water. Then turn on the disposer. Gradually feed waste into the disposer using a spatula or other long-handled utensil (don't stick the utensil in past the splash guard). Make sure your hands remain clear of danger.

To clear the drain line of food particles, run the disposer and the water for a minute or so after the food has been ground. Finally, after using the disposer, replace the drain stopper to keep objects from falling in. For tips on troubleshooting disposers, see page 42.

(Note: Your disposer will digest most types of food waste, even bones, coffee grounds, chopped up melon rinds, and small quantities of grease. About the only exception to this would be fibrous wastes such as corn husks.)

1 If you're fitting the disposer to an existing sink, start by disconnecting the trap assembly from the *basket strainer* and the drainpipe. Then remove the locknut holding the strainer in place. You may need to jar the locknut loose as shown here. Lift out the strainer and clean away any old putty from around the sink's opening.

2 To install the disposer mounting assembly, first remove the *snap ring, mounting rings,* and *gasket* from the flange. Then lay a ring of putty around the sink opening, and seat the flange in the opening. Slip the gasket, mounting rings, and snap ring up onto the neck of the flange. The snap ring will keep everything in position temporarily.

Use the slotted screws provided with the disposer to tighten the mounting assembly against the sink. Draw up the top mounting ring evenly to ensure a good seal.

3 Secure the *drain elbow* to the disposer, and if you will be draining a dishwasher through the unit (check local codes), remove the knockout inside the *nipple* above the drain elbow. (Don't forget to take the knockout out of the grinding chamber.) Now, aligning the holes in the disposer's flange with the slotted screws of the mounting assembly, lift the disposer into position. Secure it by starting the nuts onto the screws.

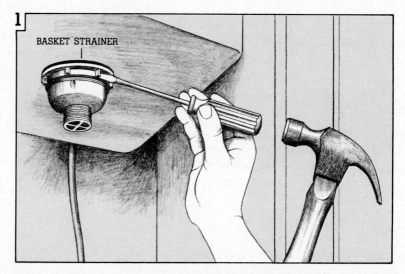

1 BASKET STRAINER

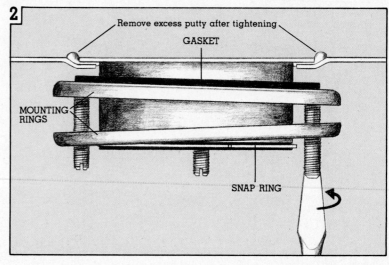

2 Remove excess putty after tightening

GASKET

MOUNTING RINGS

SNAP RING

Before tightening the nuts, rotate the disposer so the drain elbow is straight out from the drainpipe. This makes assembling the trap much easier. Tighten the nuts evenly for a good seal.

4 Fit a *slip nut* and a *washer* onto the drain elbow as shown, then fasten the trap to the elbow and the drainpipe. If necessary, saw off part of the elbow to make the connection.

If you will be draining a dishwasher through the disposer, connect its drain hose to the nipple. Use an automotive clamp for a secure connection tightening it with a standard screwdriver.

Test the installation for leaks by running water down through the disposer. If you notice any water, snug up the connections and test again.

5 Naturally, for a disposer to operate it will need power. To supply it, tap a nearby junction box and run cable to a receptacle or a junction box located below the sink. **(Be sure to shut off power to the circuit before beginning.)** If you're lucky, a box may be there already.

Now, install a switch box to one side of the sink and about 6 inches above the counter top. Run cable from this box to the junction box, and

from the disposer to the junction box. The top detail shows the hookup at the disposer; the bottom one, the connections at the junction box beneath the sink.

Restore power to the circuit, run cold water down into the disposer, turn it on, and try out your new unit.

If you'd rather not make the electrical hookup yourself, call in an electrician.

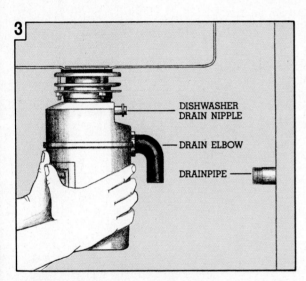

3

DISHWASHER DRAIN NIPPLE

DRAIN ELBOW

DRAINPIPE

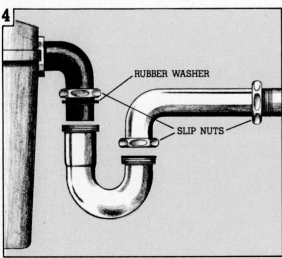

4

RUBBER WASHER

SLIP NUTS

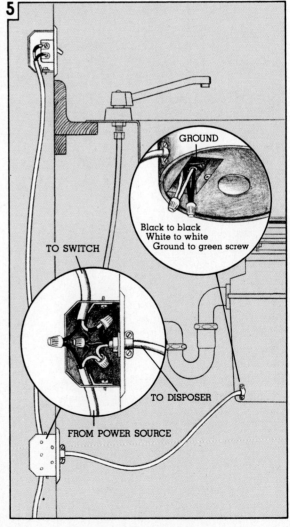

5

GROUND

Black to black
White to white
Ground to green screw

TO SWITCH

TO DISPOSER

FROM POWER SOURCE

Installing a Hot Water Dispenser

Does having an appliance that will serve up a generous supply of *very* hot water—for soups, instant coffee, tea, and so forth—seem an impossible dream? Not so! With an instant hot water dispenser, you can make quick work of food preparation chores requiring hot water, and do it less expensively than would otherwise be possible.

Here's how a hot water dispenser works. When you open the *faucet valve,* unheated water enters the unit via the water supply tube. This pressure forces hot water from the holding tank and the *expansion chamber* out the spout, which in turn reduces the temperature of the water in the tank. When it falls below a specified level, a thermostat activates *heating elements* in-

side the tank, which quickly return the water to the correct temperature.

After a hot water dispenser is installed, water may continue to drip from the spout even when the valve is in its "off" position. If this happens, lowering the thermostat setting should make things right.

Keep in mind, too, that the spout and water coming from it reach nearly boiling temperatures, so exercise care whenever you use the dispenser. And be sure to keep all flammable items away from the holding tank beneath the sink.

Finally, if you have a dispenser in a summer home, winterize it each fall by first shutting off its circuit power, then closing the water supply valve and siphoning the remaining water out of the unit.

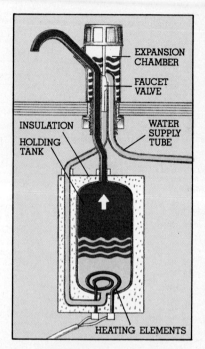

EXPANSION CHAMBER
FAUCET VALVE
INSULATION
HOLDING TANK
WATER SUPPLY TUBE
HEATING ELEMENTS

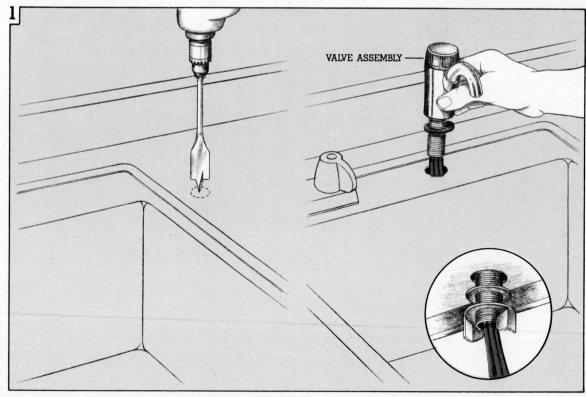

1

VALVE ASSEMBLY

2

TANK MOUNTING
BRACKET

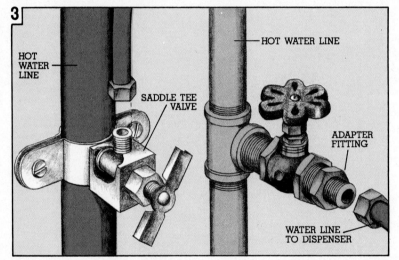

3

HOT
WATER
LINE

HOT WATER LINE

SADDLE TEE
VALVE

ADAPTER
FITTING

WATER LINE
TO DISPENSER

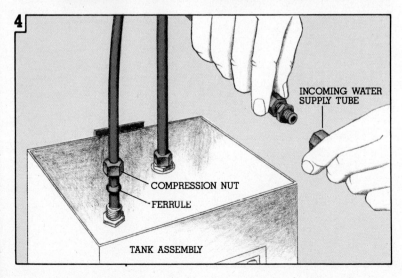

4

INCOMING WATER
SUPPLY TUBE

COMPRESSION NUT

FERRULE

TANK ASSEMBLY

1 Begin by deciding where you want to place the faucet valve assembly. In existing situations, most people prefer to drill a hole near the back edge of the counter top adjacent to the sink. However, if you're also installing a new sink, select one with an opening designed to accept one of these dispensers.

If necessary, bore a 1¼-inch hole in your counter top, then insert the valve assembly into that hole or into the one in the sink. Make your way into the compartment below the sink and secure the valve as shown in the detail.

2 Use screws to fasten the tank mounting bracket to the wall, making sure the bracket is plumb. Place it from 12 to 14 inches below the underside of the counter top. Now mount the tank onto the bracket.

3 To supply the unit with water (be sure to shut off the water and to drain the lines first), you'll need to tap into the hot water line serving the sink. If codes permit, the easy way to go is a saddle tee. To install one of these, drill a small hole in the supply line, then secure the clamp to the line as shown here. If saddle tees aren't allowed in your locale, tap in with a standard tee and install a stop valve and an adapter as shown.

4 Complete the plumbing hookup by securing the two longer tubes to the tank assembly connections, and the shorter one to the water supply tube. Note that the longer tubes are color-keyed to the appropriate connection. Restore water pressure and check for loose connections.

To energize the dispenser, plug its electrical cord into a nearby outlet. If none is handy, tap into a junction box and bring power to a new receptacle installed beneath the sink. For help running new electrical wiring, read step 5, page 59.

Installing a Hand Shower

Though somewhat of a new-comer to the plumbing scene, hand showers have caught on quickly. Not surprisingly, manufacturers responded to its popularity and now have a range of products that makes it possible to install one of these plumbing niceties in any tub or shower. There's even an adapter available that lets you add a hand shower to sink and lavatory spouts.

As you'll see below and on the opposite page, installation is neither difficult nor time-consuming. In fact, unless you encounter some unusual problem along the way, you should be able to complete the job in an hour or two.

Tub/Shower and Shower-Only Installations

1 Start by removing the existing head from the shower arm. Protect chrome parts from becoming marred by padding the jaws of the pliers or wrench you use to remove the nut securing the head. Then clean the threads of the shower arm with a wire brush or steel wool. If the arm looks like the one in the detail, you'll need to remove and replace it with one with male threads.

2 Wrap pipe tape clockwise around the shower arm threads. Then, depending on the setup you decided to go with, thread a diverter valve, a hose fitting with hanger bracket, or the hand shower hose itself onto the shower arm. The advantage of choosing a diverter valve is that you don't have to choose between having a regular *or* a hand shower. Pulling up the knob automatically re-channels the water to the hand shower. Make any other needed connections at the shower arm, again using pipe tape to seal the joints. Tighten the nuts with pliers or a wrench.

3 Connect the other end of the hose to the hand shower and if necessary mount a hanger bracket to the wall at a convenient height. If the wall is too thin to accept screws, use hollow-wall fasteners such as toggle bolts to secure the mount.

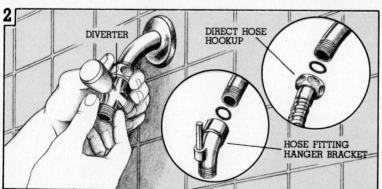

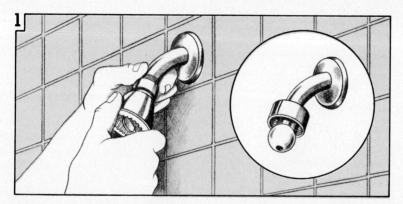

Tub-Only Installations

1 Begin your project by removing the existing tub spout. To do this, insert the handle of a hammer or other suitable prying device into the spout as shown. Once you've budged it loose, you should be able to twist it off the rest of the way with your hands. If you plan to reuse the spout somewhere else, be careful not to disfigure the chrome finish.

Now clean the threads of the spout nipple with steel wool or a wire brush. If the threads are badly rusted, apply some penetrating oil first. Allow it to soften things up, then clean the threads.

Before installing the new diverter spout, wrap the nipple threads with pipe tape. Doing this ensures a good, tight joint. It also should make it easier to remove the spout if that ever becomes necessary. Now start the spout onto the threads and tighten by hand only. Be careful that you don't damage the *hose connection* or the *lift knob* in the process.

2 With the new spout now in place, next wrap some pipe tape (clockwise) around the hose connection. Now thread the nut at the end of the shower hose onto it. Here again, take care not to exert too much pressure on the hose connection. Thread the nut at the other end of the hose onto the hand shower.

3 To complete the installation, position and mount the shower hanger. To mount most hangers you have to mark the position of the screw holes on the wall, and after drilling the holes insert anchors flush with the wall surface and secure the hanger with screws. To mount the other hanger shown here, you simply peel the paper off its back, then press it onto the wall in the desired location.

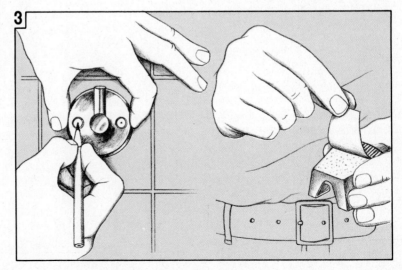

SHOWER HOSE FITTING

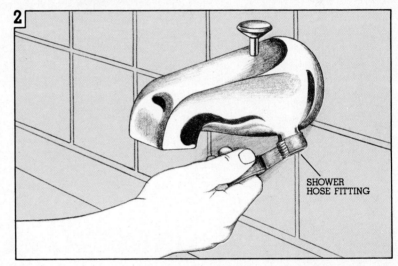

Installing a Dishwasher

Building in an automatic dishwasher is one of those projects that's easier than it looks. Most units fit neatly into a 24-inch undercounter cavity, and all are prewired and ready for simple supply, drainage, and electrical hookups.

Try to position the dishwasher as near the kitchen sink as possible. This not only simplifies the connections, but also makes sense in terms of getting dishes to the appliance from the sink.

1 Start by removing a 24-inch-wide base cabinet or otherwise tailoring a space to fit. Then bore a hole large enough to allow for passage of the supply and drain lines near the lower back of the side panel of the adjoining cabinet. To supply power to the unit, run a 15- or 20-amp circuit from the service panel. (You may want a licensed electrician to help with this part of the project.)

2 Shut off the water supply to the hot water line you plan to tap into. Drain the line, then cut into it and insert a tee fitting. Install a stop valve in the dishwasher supply line, then run flexible copper tubing from there into the cavity. This sketch shows two supply hookup alternatives. (For information on how to work with the various types of pipe, see pages 82-90.)

3 Your dishwasher can drain either into the sink drain or into a food waste disposer if you have one. For sink drainage, you'll need a dishwasher waste fitting. To install one of these, loosen the slip nuts tying the tailpiece to the trap and the sink's basket strainer. Remove the tailpiece and insert the waste fitting in to the trap. Secure it by tightening the slip

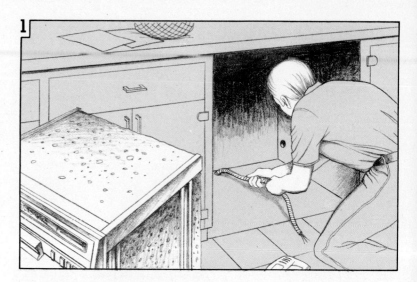

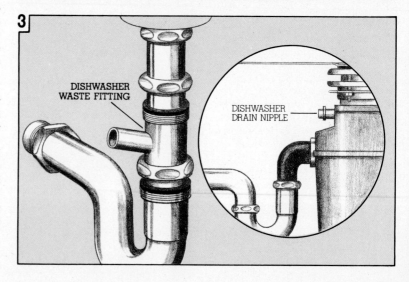

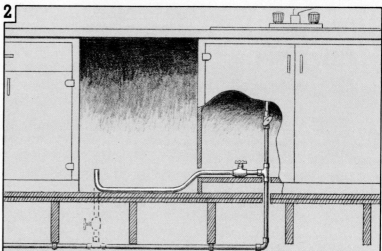

DISHWASHER WASTE FITTING

DISHWASHER DRAIN NIPPLE

4

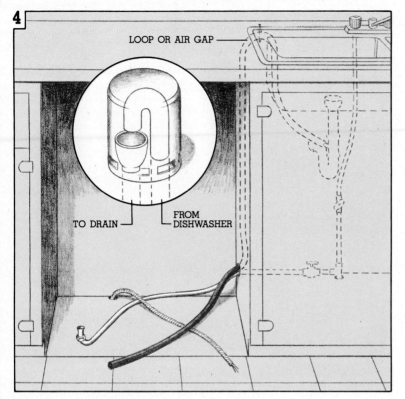

LOOP OR AIR GAP

TO DRAIN

FROM DISHWASHER

nut onto the trap. Cut the tailpiece so it is slightly longer than the distance between the basket strainer and the top of the waste fitting. Reinsert the tailpiece to complete the connection.

To drain a dishwasher into a food waste disposer, simply remove the metal knockout inside the dishwasher drain nipple on the disposer's side. Use a screwdriver and hammer for this. The knockout, when freed, will fall into the grinding chamber. Be sure to take it out.

4 Now, using an automotive hose clamp, secure the dishwasher drain hose to the newly installed waste fitting or to the disposer's dishwasher drain nipple. Thread the hose into the dishwasher cavity. To ensure proper operation of the appliance, the hose must make a loop as shown here. (Some local codes require that you install an air gap—see the detail—on the counter top instead.) Support the hose by wrapping a couple lengths of wire around it and fastening them to nails driven in the underside of the counter top.

5

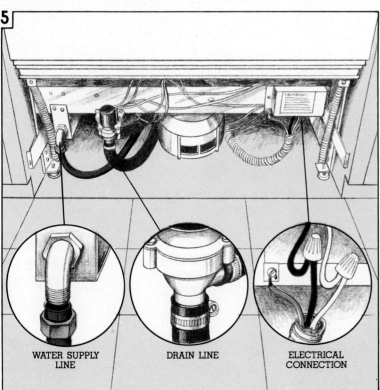

WATER SUPPLY LINE

DRAIN LINE

ELECTRICAL CONNECTION

5 The supply, drain, and electrical hookups come next. First slide the dishwasher into its cavity. Then, working your hands in under the front of the dishwasher, make the connections as shown in the details.

Complete your installation by leveling the legs of the unit and anchoring it to the underside of the counter top with the screws provided. Open the water supply line, energize the circuit supplying the electricity, and check the unit for proper operation and for leaks by running through a load of dishes. Tighten any loose connections.

Installing a Water Heater

Though it's not unheard of for water heaters to last as long as 25 years, even the best of them eventually give in to rust and corrosion. When yours does, you've only one of two choices: Call in the pros or tackle the job yourself.

Fortunately, almost anyone can learn how to install a new heater in very short order. More about that below and on the opposite page.

But first let's look at some signs of water heater failure. Probably the most easily detected is "not having enough hot water." Rust and other sediment build up in the bottom of the tank, eventually reducing the water surface area. When this energy-inefficient situation exists, the heater must work more often to satisfy demand.

Watch, too, for leaking around the base of the unit. Often, this results because the tank has rusted through. Severe rusting of the heater also indicates it's time to act.

What size unit should you buy? If you're replacing a worn-out unit, look on its nameplate and note the gallon size. Generally, you'll be safe purchasing the same size, that is unless you have recently installed or plan to buy a large user of hot water such as a clothes washer.

For new installations, 30-50-gallon units are typical. Discuss your situation with the supplier. He probably will ask you the number of bathrooms and bedrooms you have as well as the number of persons in your household, then will con-

sult a chart to determine the correct size.

Be prepared, too, to decide whether you want a *fast recovery* unit. This feature will cost you more initially and more to operate. Standard units won't heat water quite as quickly, but they are more economical.

Many of today's generation of water heaters are energy efficient. But some aren't. This, along with the quality of the material the heater is made of, accounts for the wide price differential you'll notice when shopping. Insist on one well insulated and made of glass-lined steel. You also may consider purchasing a wrap-around insulation kit to further improve the heater's efficiency.

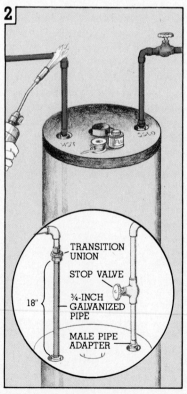

TRANSITION UNION

STOP VALVE

¾-INCH GALVANIZED PIPE

MALE PIPE ADAPTER

18"

1 Professional plumbers agree that the toughest part of installing a new water heater is unhooking and removing the old one.

Start by shutting down power to the heating elements. With gas- or oil-fired units, close the valve at the meter (or the in-line valve near the heater). If yours is an electric heater, de-energize the circuit servicing the unit.

Then, shut off the water supply at the meter, and drain your home's water lines. Next open the water heater drain valve.

After this, cut into the water lines (Mark hot and cold lines so you don't hook up the new heater backwards.) With electric heaters, remove the electrical junction box plate and disconnect the power supply wires. For gas- or oil-fired water heaters, disconnect the fuel supply pipe from the control valve. Disconnect the vent stack from the draft diverter.

Move the new unit into place; check for plumb and level, shimming if necessary.

2 How you hook up the water lines depends on the type pipe you use. Inlet (cold) and outlet (hot) connections generally are threaded. To adapt them for use with copper or plastic lines requires a special fitting.

With copper lines, apply pipe compound or pipe tape to the threads of a male adapter fitting, then screw it to the inlet or outlet connection. Now solder copper pipe to the adapter, and using the appropriate fitting connect this length of pipe to the existing lines.

If you're running plastic, connect the lines as shown in the detail. Note that the first 18 inches of the hot water line is galvanized pipe.

3 A temperature and pressure relief valve monitors conditions inside the water heater. Though the location of the valve varies from make to make, it hooks up like the one shown here.

4 Energy supply hookups come next. This sketch shows how to go about it, for both electric and gas-fired units.

5 Gas- and oil-fired units need venting to carry combustion-created gases and vapor to the outside. When connecting the flue pipe to the draft diverter (use only galvanized pipe suitable for gas venting), maintain at least a ¼-inch rise per foot for adequate draft.

6 With all connections made, open the water supply valve. After the tank and water lines have filled, bleed the lines to get rid of any air in them. Then open the fuel line stop (or energize the electrical circuit), and test for leaks. For gas- or oil-fired heaters, light the pilot according to instructions on the control valve plate and adjust the temperature setting.

3
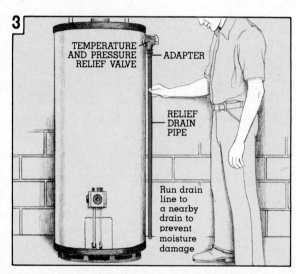
TEMPERATURE AND PRESSURE RELIEF VALVE — ADAPTER

RELIEF DRAIN PIPE

Run drain line to a nearby drain to prevent moisture damage

4
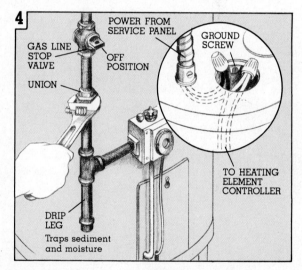
POWER FROM SERVICE PANEL

GROUND SCREW

GAS LINE STOP VALVE — OFF POSITION

UNION

TO HEATING ELEMENT CONTROLLER

DRIP LEG
Traps sediment and moisture

5
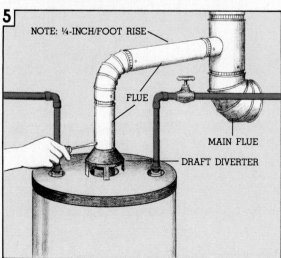
NOTE: ¼-INCH/FOOT RISE

FLUE

MAIN FLUE

DRAFT DIVERTER

6
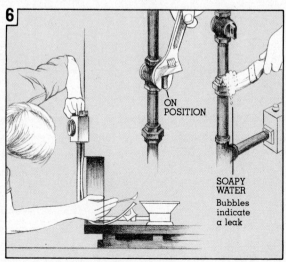
ON POSITION

SOAPY WATER
Bubbles indicate a leak

Tapping In for a New Fixture

The improvements shown on the preceding pages can be considered end-of-the-line jobs—you simply hook new fittings or fixtures to pipes that are more or less where you want them. Now let's look at several different ways you can run lines to a spot where you didn't have a fixture before.

With water, getting there is the easiest part of the problem. In fact, supply lines are the last item a plumber considers when he lays out a new run. Far trickier are the drain-waste-vent (DWV) lines that carry away water, waste, and potentially harmful sewer gas.

These and the pages that follow show only the easiest ways to tap in for a new fixture—and our hookups apply only to liquid-waste carriers such as lavatories, tubs, showers, and some sinks. For a toilet, kitchen sink, or the elaborate DWV lines you'd need for a new bathroom, you may want to consult a plumber.

Realize, too, that even for these relatively simple modifications, you'll probably need to apply to your local building department for a permit, and arrange with them to have the work inspected before you cover up any new pipes.

Planning Drain Lines

Before you begin to think about extending your plumbing system, map out exactly where its existing lines run. Your home probably has one or more drainage arrangements similar to the one shown here.

Notice that the existing fixtures, indicated in white, cluster near a *wet wall*. It's usually a few inches thicker than other walls to accommodate the 3- or 4-inch-diameter *stack* required in residential construction. (To find your wet wall, note where the stack is on your roof.) Hot and cold supply lines (not shown) may be in the wet wall, too.

The fixtures drain into the main stack directly or into *horizontal runs* that slope toward it at a pitch of at least ¼ inch per foot. Runs of more than a few feet also must be vented (check local codes), usually via a short *vertical run* and a *circuit vent* that returns to the stack. This piping is concealed inside normal-thickness walls and floors.

The fixtures shown in blue depict your main options. Generally, the closer to the stack you can get, the easier your job will be.

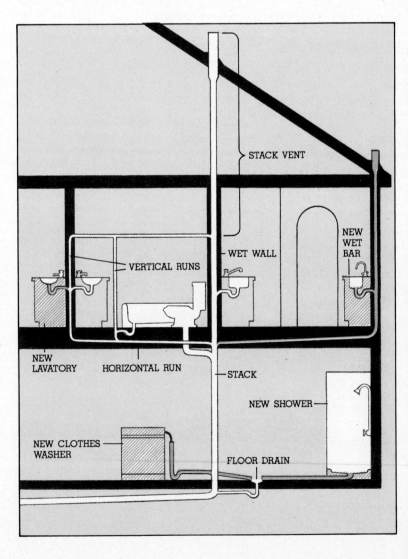

STACK VENT

NEW WET BAR

WET WALL

VERTICAL RUNS

NEW LAVATORY

HORIZONTAL RUN

STACK

NEW SHOWER

NEW CLOTHES WASHER

FLOOR DRAIN

Venting Possibilities

Think of a main stack as a two-way chimney—water and wastes go down, gases go up. Just as you wouldn't install a fireplace without a chimney, neither should you consider adding a fixture without properly venting it. Strangle a drain's air supply and you risk creating a si-phoning effect that can suck water out of traps. This in turn breaks the seal that provides protection from gas backup. The flow of wastes oftentimes becomes retarded too.

Codes are very specific about how you must vent fixtures, and these requirements differ from one locality to another. So check your community's regulations for details about the five household systems illustrated below.

Unit-venting—sometimes called *common-venting*—lets two similar fixtures share the same stack fitting. With this method you can put a new fixture back to back with one that already exists, creating a Siamese

arrangement on opposite sides of the wet wall. You simply open up the wall, replace the existing *sanitary tee* fitting with a *sanitary cross* and connect both traps to it. The drains of unit-vented fixtures must be at the same height.

Wet-venting lets a portion of one fixture's drain line also serve as the vent for another. Not all codes permit wet-venting, and those that do speci-fy that the vertical drain be one pipe-size larger than the upper fix-ture drain; in no case can it be smaller than the lower drain.

Realize, too, that regardless of how you vent a fixture, codes limit the distance between its trap outlet and the vent. These distances depend on the size of drain line you're running. For 1¼-, 1½-, and 2-inch drain lines—the sizes you'll most likely be working with—2, 3, and 5 feet are typical figures.

Thinking about installing a new shower, sink, or washing machine in your basement? If so, codes may allow you to *indirect vent* the new

fixture into a floor drain, as we did with the shower and clothes washer on the opposite page. In this situa-tion, the open drain serves as a vent for the fixture.

In a few situations, you might also be able to tap into a *circuit vent*. The new fixture must be between the vent and the stack, though. And don't try this with anything other than a small lavatory or bar sink; existing DWV lines might not be able to handle the load from a big water-user.

As a last resort, consider a *sepa-rate vent* for a fixture placed some distance from the stack. Without a doubt, this is the most costly and time-consuming . way to go, but it offers the utmost in flexibility. More about separate venting on page 75.

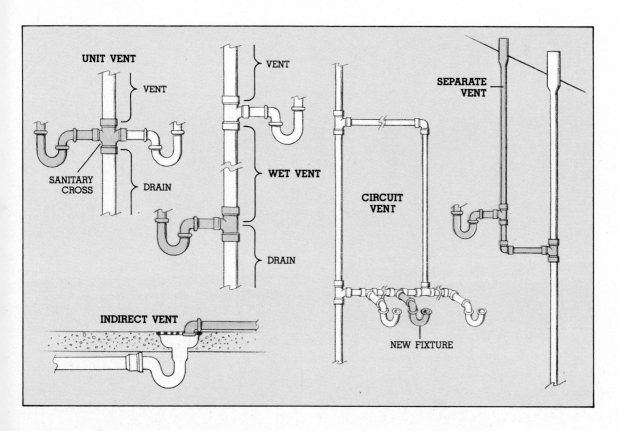

UNIT VENT · VENT · SANITARY CROSS · DRAIN · VENT · WET VENT · DRAIN · CIRCUIT VENT · NEW FIXTURE · SEPARATE VENT · INDIRECT VENT

Tapping into Wet Wall Lines

Exactly how you connect a new fixture to existing drain and supply lines depends partly on what the lines are made of and partly on the materials permitted by your plumbing code.

The drawings here show the plan of attack you'd follow in a newer home that has a plastic or copper drain-waste-vent stack and supply lines. In an older house plumbed with a cast-iron stack and galvanized supply pipes, adapt the procedures shown on pages 72–75.

Before you begin, shut off your home's main water supply. Drain the lines by opening faucets at the system's lowest point. Also provide ventilation for sewer gas that escapes when you cut into the stack. Or cap the lines with duct tape.

1 Open up the wet wall to the center of the studs on either side of the stack. This gives you plenty of room to maneuver and makes it easier to patch the opening later.

A big cutout also helps you seek out supply lines, which may or may not be in the same cavity. If not, and you don't want to make another opening, bring them up from below.

Next, firmly anchor the stack, using a pair of *riser clamps* that rest on cleats above and below the point where you'll be cutting in.

2 Now it's time for some careful measuring. Mark the fixture's rough-in dimensions on the wall, making sure the location doesn't exceed the maximum code-permitted distance explained on page 69. Once you know exactly where the trap will be, draw a line that slopes from that point to the stack at the rate of ¼ inch per foot. This tells you exactly where the new fitting's inlet opening must be located.

Cut either copper or plastic with a hacksaw. Take your time when doing this: The cuts must be square. Remove a section of pipe that is about 0 inches longer than the *sanitary tee* you'll be installing.

3 Taking out a sizable section of the stack gives you enough room to fit in the tee with a couple of *spacers* cut from the leftover stack pipe. Fit the tee, spacers, and a pair of *slip couplings* into place as shown. Slide the couplings up and down to secure the spacers.

Tubs, showers, and most sinks and lavatories commonly require that you install 1½-inch drain pipe and fittings; some lavatories hook up with 1¼-inch stock.

At this point, just dry-assemble the components. Don't solder or solvent-weld them until you've cut and fit the remainder of the drainage run.

RISER CLAMP

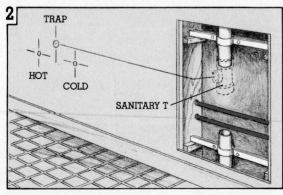

TRAP

HOT   COLD

SANITARY T

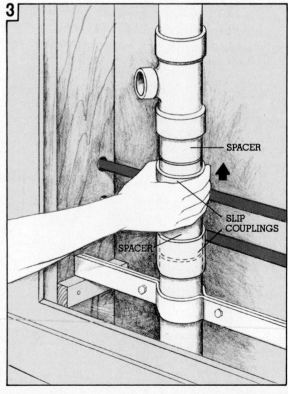

SPACER

SLIP COUPLINGS

SPACER

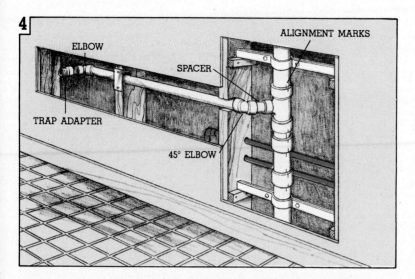

ALIGNMENT MARKS
ELBOW
SPACER
TRAP ADAPTER
45° ELBOW

4 For concealed runs, cut out a strip of drywall and notch the studs just deeply enough to support the pipes. You can secure an exposed run to the studs, using pipe straps.

You'll probably need a *45-degree elbow* and short *spacer* to negotiate the turn at the stack, and a *90-degree elbow* and a *trap adapter* at the trap.

Once you're satisfied that everything lines up properly and falls ¼ inch per foot toward the stack, scribe each pipe and fitting with *alignment marks*, then go back and permanently solder or solvent-weld each component. To learn about soldering copper pipe, see pages 82 and 83; for plastic, turn to pages 88–90.

Concealed pipes should be protected with metal plates nailed to each stud as shown. This wards off any nails or screws that might be driven into the wall after the pipe is covered.

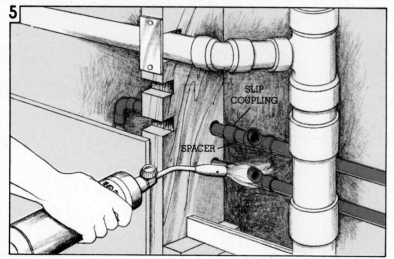

SLIP COUPLING
SPACER

5 Tap rigid copper and plastic supply lines using similar spacer/slip coupling/tee setups; with flexible tubing, you can dispense with the coupling and spacer, cutting out just enough to fit in the tees.

Again, cut, pre-assemble, and check the runs before completing the connections. Supply lines should slope away slightly from the fixture so they can be drained. If you'll be soldering copper, wet down the wall cavity and have water on hand in case nearby wood catches fire.

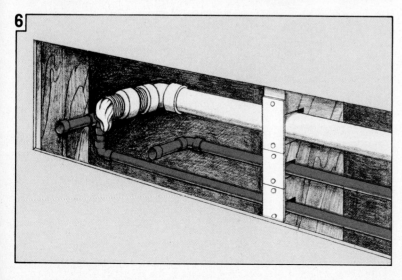

6 Unless you'll be connecting the new fixture right away, cap off the supply lines, turn on the water, and check for leaks. Cap the drain, too, or stuff a rag into it so sewer gas can't escape into your home. Don't close up the wall until you've tested all supply and drain connections. To learn about hooking up the fixture itself, see pages 45–57.

Tapping into Exposed Basement Lines

Many older homes originally were equipped with hub-type cast-iron drain pipes and galvanized supply lines. But that doesn't mean you have to stick with these materials when you add a new fixture. A no-hub fitting gets you into the stack with a minimum of hassle, and special adapter fittings let copper or plastic supply lines take over where the galvanized leaves off. Here's how to make the transition in a typical under-the-floor situation.

1 Kitchen sinks and branch drains from first-floor bathrooms often join the stack just below floor level. A tee fitting like this one offers an ideal spot to tap in.

Begin by securely supporting the stack above and below; cast iron is heavy, so brace it well. You may also need to add a hanger to hold up the horizontal run.

2 To cut into the stack you'll need to rent or borrow a chain-type pipe cutter. Wrap the chain around the stack, hook it, and, with the handles open, crank the chain tight with the turn screw. Then draw the handles together. This part of the job takes muscle, but the cutters at each link will eventually prevail and cause the pipe to break cleanly. Cut just beneath the pipe's hubbed end.

3 Make a second cut about 4 inches below the first, then cut the horizontal run. Have a helper on hand—the fitting may simply drop clear, or you may have to wrestle it loose from the hub above. If that doesn't work, hit the fitting with a heavy hammer, or melt the lead in its hub with a torch.

EXISTING DRAIN

HUB-TYPE CAST IRON STACK

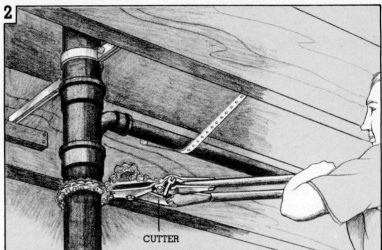

CUTTER

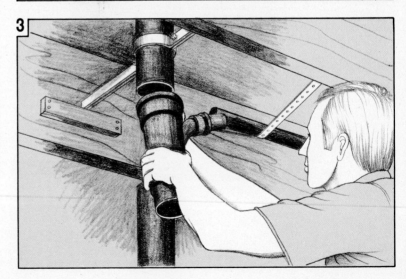

4 No-hub fittings were designed to be completely compatible with old-style cast-iron pipe. For this connection you need a *sanitary cross,* a couple of *spacers,* and clamps sized to suit the cross' inlets and outlets.

To assemble the connection, slip clamps over the pipe ends, insert the fitting, slide the clamps into place, and tighten them with a screwdriver. One advantage of no-hub is that you needn't be fussy about aligning fittings, as you must with plastic or copper. If a connection is out of kilter, you simply unscrew the clamps, straighten out the fitting, and retighten. For more about working with no-hub, see pages 91–92.

5 Two more lengths of pipe, an elbow, and a sanitary tee complete the drain run. If codes permit, you may want to go with a plastic run instead. Avoid boring through the *sole plate* if you can. If you can't, center the hole and don't make it any larger than necessary. Leave at least ⅝ inch of wood around the hole. Support the run with a strap in the basement and a riser clamp that rests atop the sole plate. Check that the horizontal pipe slopes ¼ inch per foot.

6 Sometimes, the distance required for the new run exceeds code requirements. In that event, you need to run a separate vent. Here's what's involved.

Begin in the attic. Again, if you must bore through a plate, size the hole just big enough to handle the pipe's outside diameter. For a 1½-inch drain, you'll probably need only a 1¼-inch vent; check your code.

Now mark the point where you need a hole in the roof by driving a nail up through the sheathing. Remove the shingle that the nail penetrates and bore through the sheathing. If there's a rafter in the way, you can offset the vent with a couple of 45-degree elbows. *(continued)*

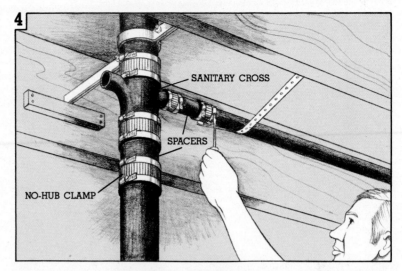

SANITARY CROSS

SPACERS

NO-HUB CLAMP

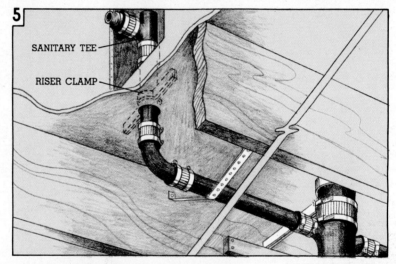

SANITARY TEE

RISER CLAMP

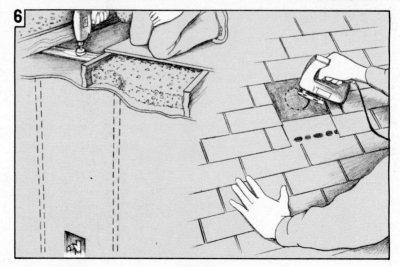

Tapping into Exposed Basement Lines *(continued)*

7 If you've used no-hub pipe to this point, you may want to consider switching to copper or plastic for the vent stack. These materials weigh much less, and the longer lengths you'll need will be easier to maneuver and set into place.

Make the switch with a *hubless adapter* at the sanitary tee. One end of this fitting accommodates no-hub clamps, the other has a socket for copper or plastic pipe.

8 Regardless of the material you decide to use for a vent, be sure to anchor it with a riser clamp to wood structural members in the attic.

In colder climates, codes generally call for an *increaser* up top where piping penetrates the roof. This prevents freeze-ups that could clog the vent. It typically measures at least 2 inches in diameter, but check your code for specifics. Also find out how high the vent must extend above the roof. One foot is typical.

Cap off your installation, using a *vent flashing* that slips over the increaser. Tuck its flange under shingles on the up-roof side and seal all around with roofing cement.

9 With the DWV system complete, all that remains is to extend supply lines to the new location. If you find a couple of convenient unions in an existing run, crack these open and dismantle them back to the nearest fittings. Otherwise, cut each supply line pipe back to the nearest fitting and turn out the pieces.

To go from galvanized to plastic or copper, use a threaded *adapter* like the one shown here. Never hook copper pipe directly to galvanized. Electrolytic action will corrode the connection.

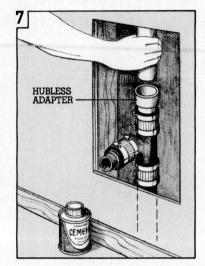

HUBLESS ADAPTER

CEMENT

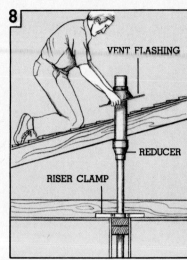

VENT FLASHING

REDUCER

RISER CLAMP

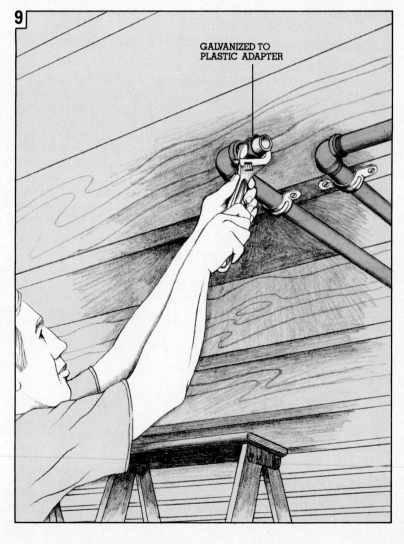

GALVANIZED TO PLASTIC ADAPTER

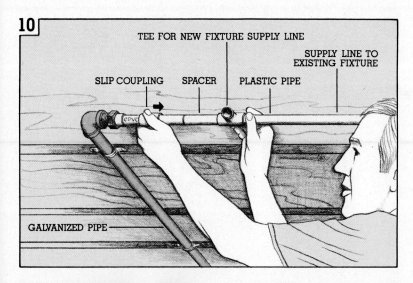

10 TEE FOR NEW FIXTURE SUPPLY LINE

SUPPLY LINE TO EXISTING FIXTURE

SLIP COUPLING   SPACER   PLASTIC PIPE

GALVANIZED PIPE

10 Now replace the run you've removed with copper or plastic pipe and a tee fitting. Splice in with a slip coupling and spacer. Don't solder or solvent-weld the connections until you've assembled the remainder of the supply line and checked to be sure everything fits properly.

Supply lines should fall slightly away from fixtures so the system can be drained easily from its lowest point. You needn't slope them as drastically as you did the drain line; just slightly off-level is adequate.

11 A pair of *drop ells* at the fixture location offers a way to both support pipes and get them around the bend. Attach them to a cleat toenailed into studs on either side of the wall cavity.

You may need to add an extra elbow to either the hot or cold supply line to space them apart the 6 or 8 inches typically required in fixture roughs.

After all dry-assembled runs check out, you can solder or solvent-weld all of its fittings, as shown on pages 82–83 and 88–90. Before you solder, open every faucet on the run. Heat from the torch can burn out washers and other parts, and built-up steam could rupture a fitting or pipe wall.

Remember, too, that plastic lines, especially those that handle hot water, need room for expansion and contraction. Enlarged holes in framing members allow for this.

The moment of truth comes, of course, when you cap off the supply lines, turn on the water, and look for leaks. Some codes require two plumbing inspections—one before you patch up the wall and install the fixture, the other when all work is completed.

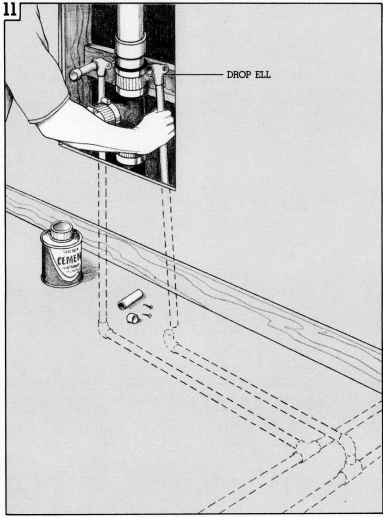

11 DROP ELL

CEMENT

PLUMBING BASICS AND PROCEDURES

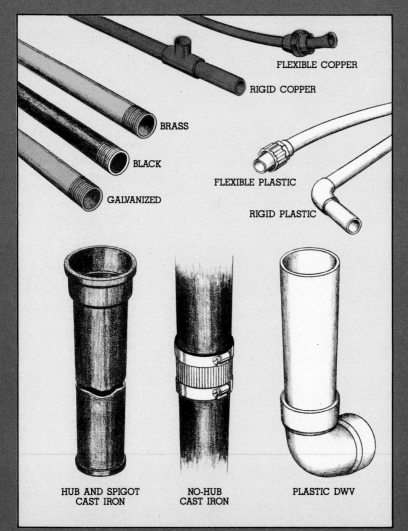

FLEXIBLE COPPER

RIGID COPPER

BRASS

BLACK

GALVANIZED

FLEXIBLE PLASTIC

RIGID PLASTIC

HUB AND SPIGOT
CAST IRON

NO-HUB
CAST IRON

PLASTIC DWV

This final section delves into the largely hidden elements of your home's plumbing system—its pipes and the fittings that join them together.

Check the drawing at left and the chart on the opposite page and you'll find you have a multitude of choices here. That's because piping materials and the systems used for assembling them have undergone a dramatic evolution in the past 40 years.

Before World War II, most homes had galvanized steel, brass, or bronze supply lines, and cast-iron drain-waste-vent (DWV) systems. After the war, copper became the favored material for supply lines, and sometimes for DWV runs as well. More recently, plastic has appeared on both the supply and DWV scenes, and no-hub clamps have made cast iron much easier to deal with.

Which you choose for a project depends partly on the job you want it to do, and partly on the materials local codes do and don't permit. You needn't, however, be limited by the existing pipes in your home. Special adapter fittings make it easy to interconnect new materials with old.

Choosing the Right Pipe Materials

Material	Type	Uses	Features and Joining Techniques
Copper	Rigid	Hot and cold supply lines; DWV*	The most widely used, although more costly than other types. Lightweight and highly durable. Sold in 20-foot (and sometimes shorter) lengths. Solder it together.
	Flexible	Hot and cold supply lines	Comes in easily bent 60- and 100-foot coils. Solder or connect with special, mechanical fittings.
Threaded	Galvanized steel	Hot and cold water lines; DWV* (not for gas)	Because it's cumbersome to work with and tends to build up lime deposits that constrict water flow, galvanized isn't widely used anymore. Standard-length pipe must be cut, threaded, and screwed into fittings, but you can buy shorter precut sections that are already threaded.
	Black steel	Gas and heating lines; vent piping	The main difference between this and galvanized is that "black pipe" rusts readily and isn't used as a carrier for household water.
	Brass and bronze	Hot and cold water lines	Again, you cut and thread. Very durable but also very costly.
Plastic	ABS	DWV* only	Black in color. Lightweight and easy to work with, it can be cut with an ordinary saw, and cemented together with a special solvent. Not all local codes permit its use.
	PVC	Cold water and DWV*	Cream colored, blue-gray, or white. PVC has the same properties as does ABS, but you can't interchange these materials or their solvents.
	CPVC	Hot and cold supply lines	White, gray, or cream colored. Same properties as ABS and PVC.
	Flexible polybutylene (PB)	Hot and cold water lines	White or cream colored. Goes together with special fittings. Costly and not widely used.
	Flexible polyethylene (PE)	Cold water and gas	Black. This material has the same properties as polybutylene. Used mainly for sprinkler systems.
Cast iron	Hub and spigot	DWV* only	Joints are packed with oakum, then sealed with molten lead. Some plumbers use a special compression-type rubber gasket.
	No-hub	DWV* only	Joins with gaskets and clamps.

*DWV is the plumbing industry abbreviation for drain-waste-vent lines.

Choosing the Right Fitting

The parts bins at a plumbing supply house contain hundreds of fittings that let you interconnect any pipe material in an almost unimaginable number of ways. To know what to ask for, though, you have to master a plumber's vocabulary that's largely alphabetical. Here are the ABCs of ells (Ls), tees (Ts), and wyes (Ys).

1 As you might guess, *supply (or pressure) fittings* can be used only in *supply lines.* You can see why by looking at the details here. A drain pipe and fitting join and form a smooth inner surface to ensure passage of solid wastes. But with supply pipes and fittings, you'll note a restriction where pipe and fitting meet.

When changing directions in a supply run, you'll need an *elbow* (ell). Most make 90- or 45-degree turns. A *street ell* has male and female connections to allow for insertion into another fitting. A *reducing ell* joins one size pipe to another.

When you order fittings, give the size first, then the material, and finally the type you want. You might, for example, ask for a "½-inch galvanized 90-degree ell." With reducing fittings, give the larger size first, then the smaller.

Tees are used wherever two runs intersect. *Reducing tees* let you connect pipes of different diameters, as you would in taking a ½-inch branch off a ¾-inch main supply line.

A *coupling* connects pipes end to end. *Reducers* let you step down from one pipe diameter to a smaller one. *Slip couplings* (not shown) get you into an existing copper or plastic line, as shown on pages 72–75.

Adapters let you go from male to female or vice-versa. Special adapters, sometimes called *transition fittings,* permit you to change from one pipe material to another.

In any run of threaded pipe, you'll need a *union* somewhere. This fitting compensates for the frustrating fact that you can't simultaneously turn pipe into fittings at either end. More about this important item on pages 86 and 87.

Caps and *plugs* close off the ends of pipes and fittings. A *bushing* lets you thread a pipe into a larger-diameter fitting.

Nipples—lengths of pipe that are less than 12 inches long—are sold in standard sizes because short pieces are difficult to cut and thread. A *close nipple* is threaded from one end to the other to join female connections.

2 Examine *drainage fittings* and note how they're designed to keep waste water always flowing downhill. Sometimes called *sanitary fittings,* they have gentle curves rather than sharp angles, where waste might get hung up.

Choose ¼ *bends* for 90-degree turns, ⅕ bends for 72-degree angles, ⅙ for 60 degrees, and ⅛ for 45 degrees.

Sanitary branches such as the *tee, wye,* and *cross* shown here come in a variety of configurations that suit situations where two or more lines converge. These can be tricky to order, so make a sketch of your proposed drainage lines, identifying all pipe sizes, and take it to your supplier when you order.

Toilet hookups require a *closet bend,* which connects to the main drain, and a *closet flange,* which fits over the bend and anchors the bowl.

Your drainage system probably has at least one *cleanout,* but consider installing another to get an auger into a drain that chronically clogs.

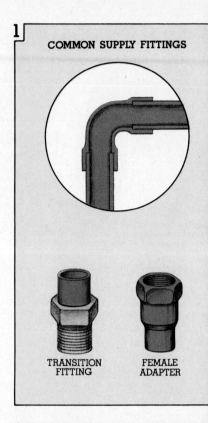

1 COMMON SUPPLY FITTINGS

TRANSITION FITTING

FEMALE ADAPTER

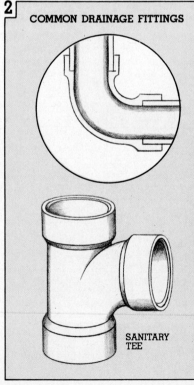

2 COMMON DRAINAGE FITTINGS

SANITARY TEE

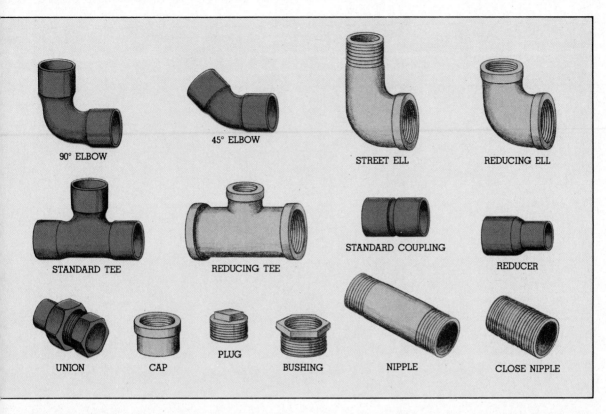

90° ELBOW

45° ELBOW

STREET ELL

REDUCING ELL

STANDARD TEE

REDUCING TEE

STANDARD COUPLING

REDUCER

UNION

CAP

PLUG

BUSHING

NIPPLE

CLOSE NIPPLE

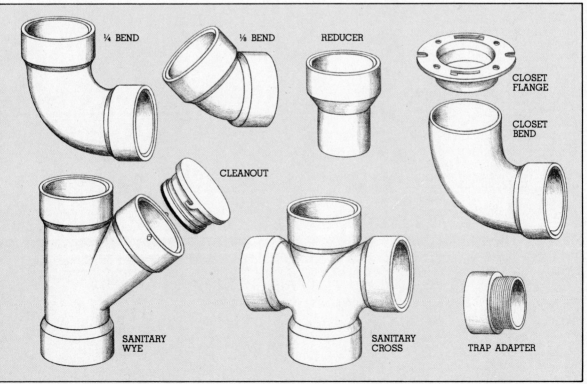

¼ BEND

⅛ BEND

REDUCER

CLOSET FLANGE

CLOSET BEND

CLEANOUT

SANITARY WYE

SANITARY CROSS

TRAP ADAPTER

Measuring Pipes and Fittings

One of the most frustrating things that can happen in a plumbing project is to shut off the water, break into a line, then discover you haven't bought the right-size pipe or fittings to put things back together again.

Errors in measuring are easy to make, too, because plumbing dimensions aren't always what they appear to be. Holding a rule to the outside of a steel pipe, for example, might seem to indicate that you're dealing with ¾-inch material. But pipes are always sized according to their *inside diameters*, as illustrated in sketch 1. Since the wall thickness of steel pipe adds another ¼ inch or so, the true size of that "¾-inch" line is actually ½ inch. This—known as the *nominal dimension*—is what you'd ask for at a plumbing supplier.

Sizing up fittings can be just as confusing. Their inside diameters must be large enough to fit over the pipe's outside diameter, so what is known as a ½-inch fitting (because it's used with ½-inch pipe) has openings that are about ¾ inch in diameter.

The third mathematical pitfall for amateur plumbers occurs when you need to compute the length of pipe necessary to get from one fitting to the next. Here you must account for the depth of each fitting's socket, as well as the distance between them.

To keep straight these critical ins and outs of plumbing measurements, study the drawings on these pages and refer to the dimensions chart for specific guidelines.

1 Measure a pipe's inside diameter if possible. This may vary from standard pipe size, depending on wall thickness. To get the nominal dimension, round off to the nearest ⅛ inch.

It's not practical to open up a pipe in an existing run just to measure it. It's much better to examine the fittings. Some manufacturers indicate on the fitting itself the size pipe it's intended for.

2 Calipers or dividers offer the most accurate way to measure outside diameters. To convert from outside to inside dimensions, see the chart opposite.

3 To figure the length of a pipe, first measure from face to face, as shown. Next, check the chart for the socket depth of the material you're working with. Since pipes have fittings on both ends, multiply by two, then add the face-to-face length.

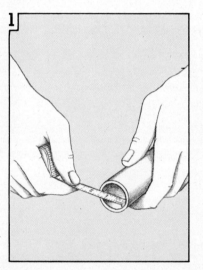

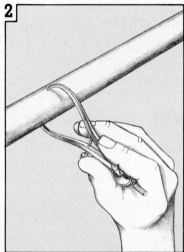

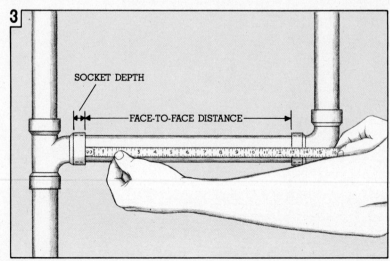

SOCKET DEPTH

FACE-TO-FACE DISTANCE

4 Pipes must engage fully in fixture sockets. Otherwise, the joint could leak. Socket depths vary from one pipe size and material to another. Only with no-hub cast-iron pipe do you not have to factor in the sockets.

When you're buying fittings, invest in a handful of different size caps. They're available in copper, plastic, and threaded. Then if you've mis-read a dimension—as even experienced plumbers do occasionally—you can cap off the line and turn the water on again before heading off to the plumbing supplier.

Refer to the chart below when you want to determine a pipe's nominal size by measuring its outside diameter, and when you need to factor in socket depths in computing lengths.

As a rule of thumb, the outside diameter (OD) of copper is ⅛ inch greater than its inside diameter (ID) nominal size; for threaded and cast iron, the figure is ¼ inch; and for plastic pipe, figure ⅜ inch to account for the wall thickness.

Copper, cast-iron, and smaller threaded fittings have standardized socket depths. With sizes ½ inch and larger, variations in tapping and threading equipment may add or subtract ¹⁄₁₆ inch or so from our dimensions. If you're planning a run with more than just a few fittings, double-check by measuring a pipe, twisting a fitting onto it, and measuring again.

Socket depths for plastic pipe vary somewhat, too, depending on the brand and whether you're dealing with ABS, PVC, or CPVC. Just place a measuring tape into the sockets of larger-diameter plastic fittings. With smaller sizes, measure the pipe, apply cement, push on a fitting, and measure again. Don't just dry-fit the connection—the pipe won't seat completely in the socket until you soften it with the cement.

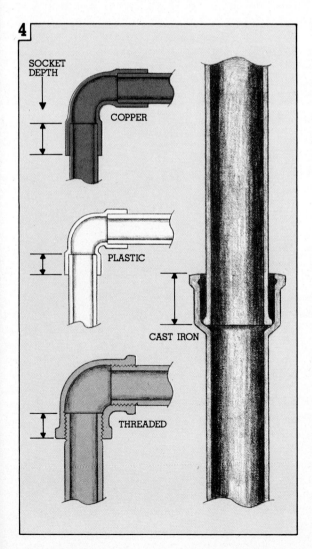

4

SOCKET DEPTH

COPPER

PLASTIC

CAST IRON

THREADED

Sizing Up Pipe Dimensions

Material	Nominal Size (approx. inside diameter)	Approx. Outside Diameter	Approx. Socket Depth
Copper	¼ in.	⅜ in.	⁵⁄₁₆ in.
	⅜ in.	½ in.	⅜ in.
	½ in.	⅝ in.	½ in.
	¾ in.	⅞ in.	¾ in.
	1 in.	1⅛ in.	¹⁵⁄₁₆ in.
	1¼ in.	1⅜ in.	1 in.
	1½ in.	1⅝ in.	1⅛ in.
Threaded	⅛ in.	⅜ in.	¼ in.
	¼ in.	½ in.	⅜ in.
	⅜ in.	⅝ in.	⅜ in.
	½ in.	¾ in.	½ in.
	¾ in.	1 in.	⁹⁄₁₆ in.
	1 in.	1¼ in.	¹¹⁄₁₆ in.
	1¼ in.	1½ in.	¹¹⁄₁₆ in.
	1½ in.	1¾ in.	¹¹⁄₁₆ in.
	2 in.	2¼ in.	¾ in.
Plastic	½ in.	⅞ in.	½ in.
	¾ in.	1⅛ in.	⅝ in.
	1 in.	1⅜ in.	¾ in.
	1¼ in.	1⅝ in.	¹¹⁄₁₆ in.
	1½ in.	1⅞ in.	¹¹⁄₁₆ in.
	2 in.	2⅜ in.	¾ in.
	3 in.	3⅜ in.	1½ in.
	4 in.	3⅜ in.	1¾ in.
Cast iron	2 in.	2¼ in.	2½ in.
	3 in.	3¼ in.	2¾ in.
	4 in.	4¼ in.	3 in.
	5 in.	5¼ in.	3 in.
	6 in.	6¼ in.	3 in.

Working with Rigid Copper Pipe

Soldering or "sweating" together rigid copper plumbing lines isn't nearly as hot and exhausting a procedure as it may sound. It's the pipes and fittings, not the installer, that do the sweating.

The term refers to a process called *capillary action*, which occurs when you heat up a joint and touch solder to it. Just as an ink blotter soaks up ink, a soldered joint absorbs molten metal, making a watertight bond that's as strong or stronger than the pipe itself.

To solder copper you need a spool of 50 percent tin/50 percent lead solder, a pastelike *flux* that both cleans the copper and helps solder flow more readily, and a propane torch.

When you shop for pipe and fittings for your first copper plumbing run, buy a few extra fittings and use them to practice the techniques shown here. Doing this greatly increases your chances of getting leak-proof joints. Leaks that show up when you fill the lines with water mean you have to drain and completely dry out the line before you can resolder the joint. More about this on the next page.

1 Cut copper pipe, using an inexpensive tubing cutter, like the one shown here, or a hacksaw. With a cutter, simply clamp the device onto the tubing, rotate a few revolutions, tighten, and rotate some more.

The *cutting wheel* makes a neat, square cut, but leaves a liplike burr inside that could restrict water flow. Remove this by inserting the *reaming blade* and twisting.

Make hacksaw cuts in a miter box and use a file to remove any burrs. Take care that you don't nick the metal, which could cause the connection to leak.

2 Now polish the outside of the pipe and the inside of the fitting with sandpaper or steel wool. This removes grease, dirt, and oxidation that could impede the flow of solder. Do a thorough job, but stop polishing when the metal is shiny. Avoid touching polished surfaces; oil from your fingers could interfere with the capillary action when you apply solder to join the pieces.

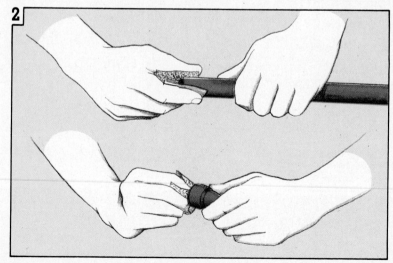

CUTTING WHEEL

REAMING BLADE

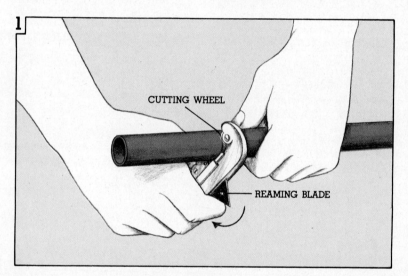

3 Apply a light, even coating of flux to both surfaces to retard oxidation when you heat the copper. As solder flows into the joint, the flux burns away. Use rosin- (not acid-) type flux for plumbing work.

4 Assemble the connection and heat the joint, using the inner cone of the torch's flame. Test for temperature by touching the joint—but not the flame—with solder as shown.

If it melts, apply more. Capillary action will pull solder around the pipe and into the joint. When molten metal drips from the bottom, remove the flame and inspect your work. A well-soldered joint has an even bead around its entire circumference. Any gaps probably will leak water.

As you practice soldering, aim to apply only enough heat to melt the solder, and only enough solder to fill the joint. Too much of either will melt all the flux and weaken the bond.

If you do have to resolder a connection, your biggest problem is to get rid of any moisture that remains inside the line. Plumbers sometimes do this by stuffing in a piece of bread (but not the crust) just upstream from the connection. This flushes easily from the system when pressure is restored and a faucet turned on. You also can buy waxy capsules that plug up the line while you work. Later, just apply heat to the point where the capsule has lodged and the capsule melts away.

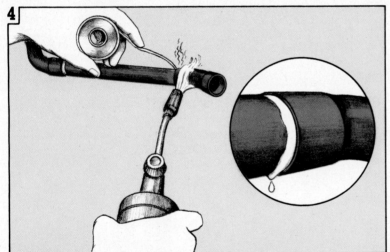

5 For a tidy, professional look, lightly brush the joint, using a damp rag. Take care that you don't burn your fingers.

Most pros lay out an entire run of copper, first cutting and dry-fitting all of its components. After dry-fitting, they go back to clean, flux, and solder each joint. Support copper pipe at 6-foot intervals.

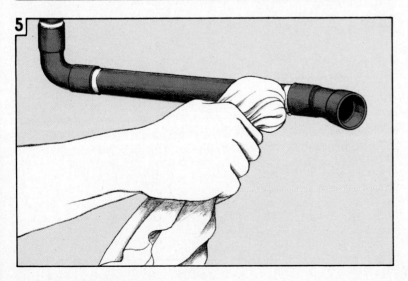

Working with Flexible Copper Tubing

Flexible copper tubing—also known as "soft copper"—is pliable enough to negotiate without elbows all but the sharpest bends. And that's good news for you. This means you don't have to install a fitting every time a run makes a change of direction, as you must when working with rigid pipe.

And though flexible accepts the same soldered fittings used with rigid, you can also make connections using the compression and flare fittings shown on this and the following page. Realize, though, that these specialized devices cost quite a bit more than standard fittings, and they aren't quite as strong, either. For economy and durability, use solder connections that will never need to be broken; save compression and flare fittings for semipermanent hookups and for those tight-quarter situations where a torch might start a fire.

Because flexible tubing isn't as damage-resistant as rigid, don't use it for long runs across open basement ceilings or in other exposed locations. For fixture connections you can buy short lengths of chrome-plated soft copper tubing. More about this on pages 48-49.

1 Cut flexible, using a hacksaw or tubing cutter, and remove any burrs as explained on page 82. When you're cutting, take care that you don't flatten out the tube ends; watertight connections demand that they be perfectly round.

Because this material is soft, always handle it gently. Uncoil tubing by straightening it out every few inches, and bend it in gradual, sweeping arcs. Otherwise, flexible will kink and you'll have to throw it away. Kinks, which impede water flow, are almost impossible to reshape.

A *spring bender*, like the one shown here, reinforces tubing walls and minimizes kinking. To use one of these, you slip the spring over the tubing and bend at several points along the arc's radius. Filling the tubing with sand also helps avoid kinking when bending it.

2 Compression fittings assemble without special tools. You just slide a *compression nut* and *ring* onto the tubing, insert the tubing and ring into the fitting, and thread on the nut.

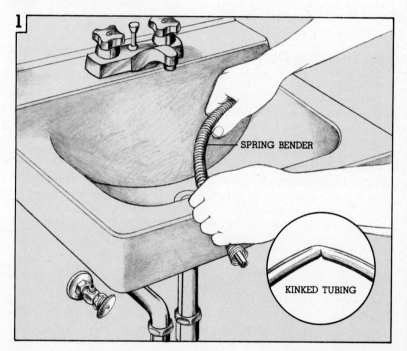

SPRING BENDER

KINKED TUBING

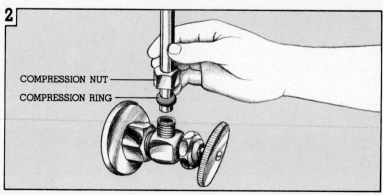

COMPRESSION NUT

COMPRESSION RING

3 Tightening the compression nut forces the ring down onto the tubing to secure and seal the connection. Be careful, though, that you don't over-tighten, which could crush the tubing and cause a leak. Go just about a quarter-turn with a wrench. If the joint leaks when you turn on the water, further tighten the nut a quarter-turn at a time.

4 For a flare fitting, you need to shape the tubing end, using either the flaring tool shown here or a tapered device that you drive into the tubing with a hammer.

The trick with both of these is to remember that you must always slip on a flaring nut before you flare the end of the tubing. With a flaring tool, you then clamp the tubing into a *block* with beveled holes sized to hold several typical pipe diameters. Align the *compression cone* in the tubing end and tighten the screw. As you turn the cone into the tubing, it flares the end. Inspect your work carefully after removing the tubing from the block. If you notice that the end has split, cut off the flared portion and repeat the process.

5 As with compression fittings, don't over-tighten a flared joint. For starters, hand-tighten, then go just a quarter-turn with a wrench. If water drips from the fitting, tighten it only until the leaking stops.

When a compression or flare fitting leaks, and gentle wrench work won't solve the problem, dismantle the joint. Was the pipe cut squarely? Were the fitting's threads meshing properly? If any parts are damaged, you'll have to start over again.

Protect compression and flare connections from strain by supporting pipe runs on either side. Since flexible tubing tends to sag when it's full of water, support horizontal lines at 4-foot intervals.

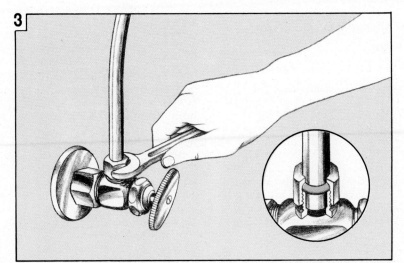

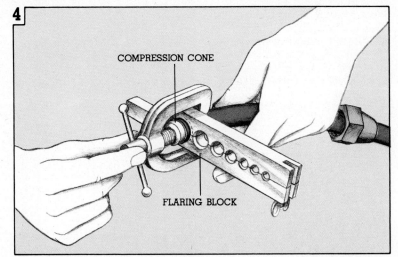

COMPRESSION CONE

FLARING BLOCK

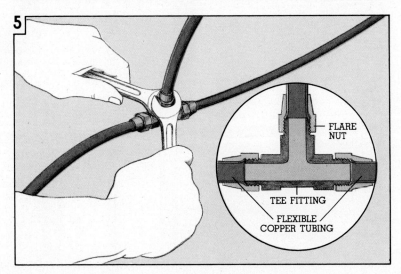

FLARE NUT

TEE FITTING

FLEXIBLE COPPER TUBING

Working with Threaded Pipe

After your first experience with threaded pipe, you'll begin to appreciate one reason why this heavy, cantankerous stuff is all but extinct in modern-day plumbing systems. Cutting, threading, and assembling it are muscle jobs—and taking apart a run that seems to have welded itself solid calls for even more brute strength.

If your home was built before World War II, probably most all of its pipes are threaded. This doesn't mean, though, that you have to stick with the same material for repair or improvement jobs. Special fittings let you break into a line of threaded, and continue on from there with plastic or copper. More about this on pages 72–75.

Threaded materials vary. Galvanized steel—recognizable by its aluminum gray color carries fresh water. A generation ago it was standard for hot and cold lines. But because it is susceptible to corrosion and buildups of lime deposits that choke water flow, galvanized is just about totally obsolete.

Black-iron pipe costs less than galvanized, but you can use it for gas or heating lines only. Its tendency to rust makes it unsuitable for household water piping.

If your home has brass or bronze piping, count yourself fortunate. These are the most durable of all (and the most expensive).

1 Examine the way pipes and fittings thread together and you'll see that you can't simply begin *un*threading them anywhere. Somewhere in every pipe run is a *union* like the one shown here. This serves as a key-stone that lets you dismantle the piping around it.

To crack open a union, examine it closely, and determine which of the smaller *union nuts* the *ring nut* is threaded onto. With one wrench on each, turn the ring nut counterclockwise. Once it's unthreaded, you have the break you need.

No union handy? Then you'll need a coarse-tooth hacksaw and some elbow grease. When you reassemble the run, don't forget that you'll need to install a union.

2 Recalcitrant old waterpipe fittings may respond to penetrating oil. If not, try heating them with a torch. Always use two wrenches—one on the pipe, the other on the fitting. And make sure that both have a good grip before you exert pressure; if a wrench slips, you could be injured.

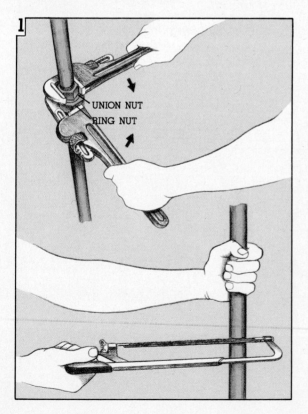
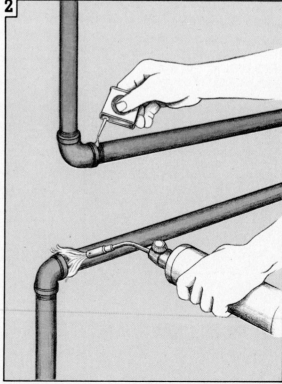

UNION NUT
RING NUT

3 If you only need a few lengths of threaded pipe, your best bet is to measure carefully (don't forget to include the depth of each fitting's socket), take the dimensions to a plumbing supplier, and let him do the cutting and threading for you.

Otherwise, you'll have to rent or borrow several pieces of specialized equipment. For cutting, get a bigger version of the device shown and explained on page 82. After a pipe has been cut, its inner lip must be pared away with a *reamer* like the one shown here. Unless you have a bench vise with *pipe jaws,* you'll also need a portable pipe vise.

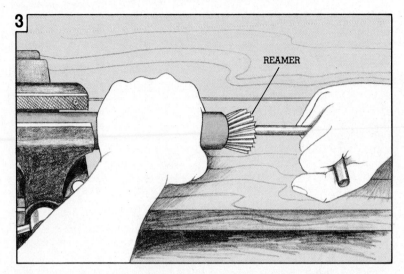

4 Two more items—the *die* and *die stock* shown here—cut the threads. Lock the proper diameter die into the stock and fit it onto the pipe end. Make sure the die starts out perfectly square or your threads won't mesh with standard fittings.

Now begin rotating the die stock. Go slowly, applying cutting oil freely as you turn. If you encounter resistance, a chip probably has jammed the threads; back up, brush away the chip, re-oil the pipe and the die, and proceed. Stop when the end of the pipe is flush with the end of the die.

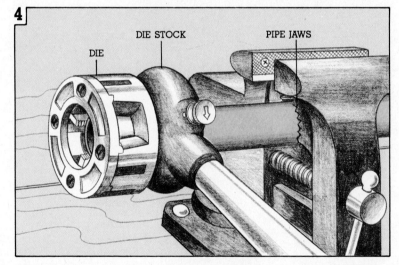

5 Before you thread a pipe and fitting together, seal the pipe's threads, using *pipe joint compound* or a special synthetic *pipe tape.* Use the vise to preassemble as much of the run as you can, making sure that pipes are turned completely into each fitting.

When you can't use the vise, always hold the fitting with a second wrench so torque doesn't break something down the line. Support runs of threaded pipe at least every 6 to 8 feet.

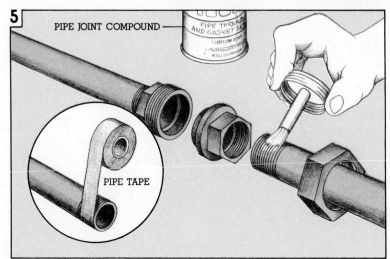

Working with Rigid Plastic Pipe

Plastic plumbing components may be the greatest thing that's happened to amateur plumbers since the invention of running water. This material cuts with an ordinary handsaw, and when it's time to assemble a run, you can forget about a torch, solder, or wrenches—everything simply cements together like the parts of a model airplane.

Not all codes allow plastic, though, and those that do spec-ify which types you can use for drainage and for cold and hot water lines. For drainage lines, your code may specify ABS only, PVC only, or either; it'll certainly require CPVC for hot-water runs.

Bear in mind, too, that you can't mix these materials. Each expands at a different rate, and there's even a specially formulated cement for each.

Expansion can cause an im-properly installed run of plastic pipe to creak, squeak, groan, and maybe even leak every time you turn on the water. To prevent this from happening allow lots of clearance between fittings and framing, and be sure to make oversize holes wherever pipes pass through a wall or floor.

Rigid plastic pipe, not surpris-ingly, isn't quite as stiff as its metal counterparts. Be sure to support horizontal runs every 4 to 5 feet.

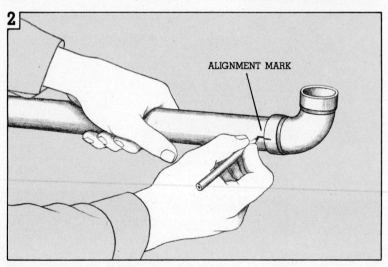

ALIGNMENT MARK

1 You can make cuts, using just about any fine-tooth saw, but use a miter box to keep them square. Diag-onal cuts reduce the bonding area at the fitting's shoulder—the most criti-cal part of the joint.

After you've made the cut, use a knife or file to remove any burrs from the inside or outside of the cut end. Burrs can scrape away cement and weaken the bond. Use a clean rag to wipe both the pipe and fitting.

2 Now dry-fit the connection by in-serting the pipe into the fitting. You should be able to push it in at least one-third of the way. If the pipe bottoms out and feels loose, try another fitting. Unlike copper compo-nents, plastic systems are designed for a snug "interference fit." Tapered walls on the inside of the socket should make contact well before the pipe reaches the socket's shoulder.

Plastic pipe cement sets up in about a minute, which doesn't give you much time to make adjustments. Scoring the pipe and fitting with an *alignment mark* like this saves twist-ing the connection back and forth after you've applied the cement.

3 With PVC and CPVC plastics, coat the inside of the socket and outside of the pipe with a special primer before applying the cement. This begins the softening-up process. ABS doesn't require a primer.

Immediately after you've primed, brush a smooth coating of cement first onto the pipe end, then into the fitting socket, and again onto the pipe. Use the cement that's right for the material you're working with, and don't let it puddle in the fitting.

4 Now push the two together with enough force to bottom out the pipe end in the fitting socket. Twisting about a quarter-turn as you push helps spread the cement evenly. Hold the pipe and fitting together a minute or two while they fuse into a single piece. You should end up with a smooth, even fillet all the way around the joint. Wipe off any excess cement.

5 If you misalign a connection, saw it off, making sure to cut squarely; then install a new fitting with a *spacer* and *slip coupling* as shown. Solvent-welded joints are strong enough to handle within 15 minutes, but don't run water through the line for about 2 hours, if possible.

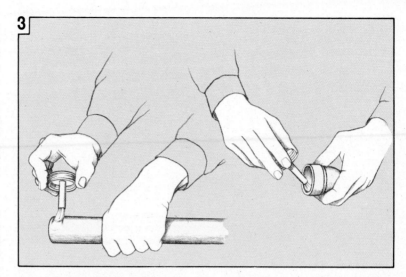

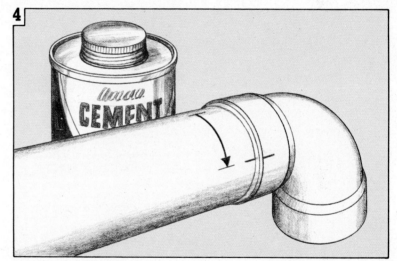

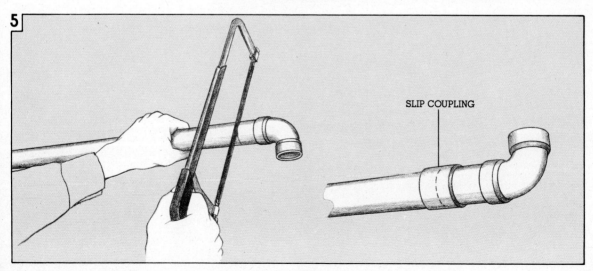

SLIP COUPLING

Working with Flexible Plastic Tubing

Think of this one as a hybrid between soft copper tubing and rigid plastic pipe. Like its copper cousin, flexible plastic comes in long coils that you can snake through tight spots without using fittings; as with rigid plastic, you have to choose the proper type for the application you have in mind, and codes in your area may disallow the material entirely.

Type PE carries cold water only, and in many communities you can use it only underground—for wells, sprinkler systems, and natural gas—where its flexibility withstands freezing and ground heaving.

Indoors, install type PB flexible plastic for hot and cold supply lines. Unlike the less-expensive type PE, it's not affected by heat.

Both types are quite easy to work with. You cut them with a knife, and join sections, using fittings like the ones shown below. You also can interconnect flexible plastic with other piping materials via transition fittings that compensate for differing rates of expansion and contraction.

Though slightly more crush-resistant than soft copper, you shouldn't use PE or PB for long, out-in-the-open runs, and it should be supported every 32 inches. Be sure to clamp it loosely so the material has sufficient room to move.

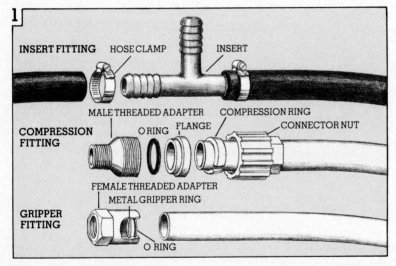

INSERT FITTING   HOSE CLAMP   INSERT

COMPRESSION FITTING

MALE THREADED ADAPTER   COMPRESSION RING
O RING   FLANGE   CONNECTOR NUT

GRIPPER FITTING

FEMALE THREADED ADAPTER
METAL GRIPPER RING

O RING

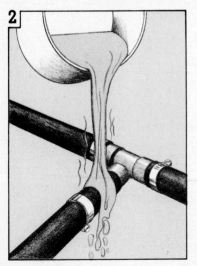

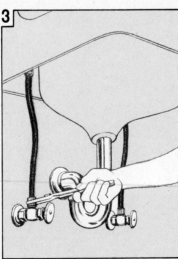

1 PE connects with plastic *insert fittings,* or you can flare the material and use the same brass fittings illustrated on page 85. With an insert fitting, slip *hose clamps* over the tubing ends, shove the tubing onto the *insert,* and tighten the clamps. Just make sure the serrations are fully inserted and the clamp is positioned squarely over top of them.

With PB, select either a *compression* or a *gripper* fitting. The ones shown connect to female and male threaded stock. You also can buy tees, ells, and other typical supply fittings.

2 Sometimes it takes lots of muscle to shove PE tubing over an insert fitting. Dousing with hot water softens the tubing so it will slip on effortlessly. Use the same technique to dismantle a stubborn connection.

3 PB makes an excellent choice for stop-to-fixture hookups. It's not quite as prone to kinking as copper is, but don't overbend. You also can buy plastic stop valves for PB and CPVC.

Working with Cast-Iron Pipe

Cast iron easily wins the heavyweight title among plumbing components. Not only is it far heftier than most materials, it also lasts longer.

You'll appreciate just how tough iron is when you first begin to wrestle with it. A 5-foot length of 3-inch pipe can weigh as much as 50 pounds. It's not easy to cut cast iron, either. And unless you go with the no-hub clamp system shown on the following page,

joining sections and installing fittings can be tricky work.

For these reasons, you may prefer to contract a plumber for a major project such as a new soil stack. He'll have the tools and know-how to do the job quickly and safely.

If you're a moderately ambitious amateur—and local codes permit no-hub—you can install smaller 1½- and 2-inch cast-iron lines yourself. Plan runs that have a minimum of bends

and that pitch back to the soil stack at a rate of ¼ inch per foot.

One of the joys of working with no-hub pipe is that it's compatible with the other cast-iron systems. This means that you can use it to tap into existing cast-iron lines, as shown on pages 72–75.

1 The easiest way to deal with the cutting problem is to measure carefully and ask your supplier to cut the lengths you need. In no-hub runs you needn't account for the size of fittings.

If you decide to do the work yourself and have more than a few cuts to make, rent a pipe cutter designed especially for cast iron. Types vary, but all save a lot of effort.

Need to make just a couple of cuts? Try fracturing the iron, as shown here. First mark all the way around the pipe's circumference with a grease pencil. Next, use a coarse-tooth hacksaw blade to cut a ¹⁄₁₆- to ⅛-inch-deep groove.

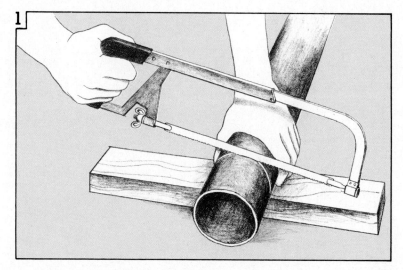

2 Now roll the pipe and whack it hard with a baby sledge. After several blows, the pipe should break cleanly. With heavy pipe, you may need to chisel around the groove to fracture the metal.

Also use the chisel to even up any jagged edges. Further smooth them by peening with the sledge or by filing. *(continued)*

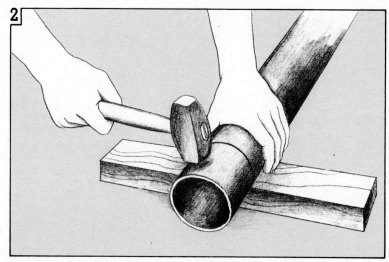

Working with Cast-Iron Pipe *(continued)*

3 Does your home have cast-iron drain-waste-vent pipes? If so, they're joined in one of the three ways depicted here. *Hub-and-spigot* piping has a bell-shaped *hub* at one end, a ridged *spigot* at the other. After inserting the spigot into the hub, a plumber packs the joint with *oakum*—a hemp saturated with pitch—then ladles in molten *lead*. Hub-and-spigot work is tricky, so leave it for the pros. You can easily break one open, though. Just chip away the lead, using a hammer and chisel, then pull the spigot free.

Piping designed for *compression* joints has hubs, but no spigots. Instead of lead and oakum, they're sealed with a special *neoprene gasket*, which is lubricated, and inserted into the hub. The spigot then is forced into place. This can be done with either a spade or a special pulling tool. Use this system to mate plastic and cast-iron piping.

No-hub joints are the easiest of all to make, and the only ones an amateur should consider. You slip a *neoprene sleeve* onto one pipe end, a *stainless steel band* onto the other, butt the pipe ends against a ridge inside the sleeve, then tighten the band's two *clamps*. No-hub pipes have the same outside diameters as those used for the other systems, so you can interconnect them easily.

4 Cast-iron pipe is heavy, so support it well. On horizontal runs, codes typically call for a *hanger* at each joint. Support a vertical run with a *riser clamp* at each floor level.

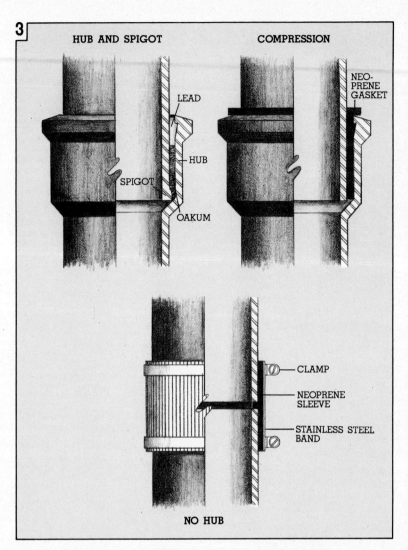

3

HUB AND SPIGOT   COMPRESSION

NEOPRENE GASKET

LEAD

HUB

SPIGOT

OAKUM

CLAMP

NEOPRENE SLEEVE

STAINLESS STEEL BAND

NO HUB

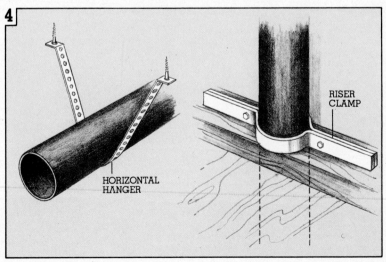

4

RISER CLAMP

HORIZONTAL HANGER

Glossary

If you are occasionally puzzled by plumbing terminology, these definitions should help. For words not listed here, or for more about those that are, refer to the index (pages 95-96).

Access panel—A removable panel in a wall or ceiling that permits repair or replacement of concealed items such as faucet bodies.

Adapter—A fitting that makes it possible to go from male endings to female endings or vice-versa. *Transition* adapters allow for joining different kinds of pipe together in the same run. *Trap* adapters help connect drainlines to a sink or lavatory trap.

Aerator—A device screwed to the spout outlet of most lavatories and sinks that mixes air with the water, which results in less water splash and smoother flow.

Air chamber—A device that provides a cushion of air for water to bang up against.

Auger—A flexible metal cable fished into traps and drainlines to dislodge obstructions.

Ballcock—The assembly inside a toilet tank that when activated releases water into the bowl to start the flushing action, then prepares the toilet for subsequent flushes. Also called a *flush valve.*

Capillary action—The action that occurs when a liquid is drawn into a razor-thin space between two almost-touching solid surfaces, such as when molten solder is drawn in and around a copper joint.

Cleanout—A removable plug in a trap or a drainpipe that allows easier access to blockages inside.

Closet bend—The elbow-shaped fitting beneath toilets that carries waste to the main drain.

Codes—See *Uniform Plumbing Code.*

Coupling—A fitting used to connect two lengths of pipe in a straight run.

Drain-waste-vent (DWV) system—The network of pipes and fittings that carries liquid and solid wastes out of a building and to a public sewer, a septic tank, or a cesspool, and allows for the passage of sewer gases to the outside.

Elbow—A fitting used to change the direction of a water supply line. Also known as an *ell. Bends* do the same thing with drain-waste-vent lines.

Fall—A word used to express the slope drain lines are installed at to ensure proper waste drainage. Minimum fall per foot is ¼ inch.

Fitting—Any connector (except a valve) that lets you join pipes of similar or dissimilar size or material in straight runs or at an angle.

Fixture—Any of several devices that provide a supply of water or sanitary disposal of liquid or solid wastes. Tubs, showers, sinks, lavatories, and toilets are typical examples.

Fixture drain—The drainpipe and trap leading from a fixture to the main drain.

Flux—A stiff jelly brushed or smeared on the surfaces of copper and brass pipes and fittings before joining them to assist in the cleaning and bonding processes.

Force cup—A suction-action tool used to dislodge obstructions from drain lines. Also called a *plunger* and a *plumber's friend.*

I.D.—The abbreviation for *inside diameter.* All plumbing pipes are sized according to their inside diameter. See also *O.D.*

Increaser—A fitting used to enlarge a vent stack as it passes through the roof.

Main drain—That portion of the drainage system between the fixture drains and the sewer drain. See also *fixture drain* and *sewer drain.*

Nipple—A 12-inch or shorter pipe that has threads on both ends and that is used to join fittings. A *close nipple* has threads that run from both ends to the center.

No-hub pipe—A type of cast-iron pipe designed for use by do-it-yourselfers. Pipes and fittings are joined using stainless steel clamps with rubber gaskets.

Nominal size—The designated dimension of a pipe or fitting. It varies slightly from the *actual size.*

O.D.—The abbreviation for *outside diameter.* See also *I.D.*

O ring—A round rubber washer used to create a watertight seal, chiefly around valve stems.

Packing—An asbestos material (used chiefly around faucet stems) that when compressed results in a watertight seal.

Pipe joint compound—A material applied to pipe threads to ensure a watertight or airtight seal. Also called *pipe dope.* See also *pipe tape.*

Pipe tape—A synthetic material wrapped around pipe threads to seal a joint. See also *pipe joint compound.*

Plumber's putty—A dough-like material used as a sealer. Often a bead of it is around the underside of water closets and deck-mount sinks and lavatories.

Plunger—See *force cup.*

P.S.I.—The abbreviation for pounds per square inch. Water pressure is rated at so many psi's.

Glossary *(continued)*

Reducer—A fitting with different size openings at either end used to go from a larger to a smaller pipe.

Relief valve—A device designed to open if it senses excess temperature or pressure.

Rough-in—The early stages of a plumbing project during which supply and drain-waste-vent lines are run to their destinations. All work done after the rough-in—setting the fixtures and so forth—is *finish work.*

Run—Any length of pipe or pipes and fittings going in a straight line.

Saddle tee—A fitting used to tap into a water line without having to break the line apart. Some local codes prohibit its use.

Sanitary fitting—Any of several connectors used to join drain-waste-vent lines. Their design helps direct wastes downward.

Sanitary sewer—An underground drainage network that carries liquid and solid wastes to a treatment plant. See also *storm sewer.*

Septic tank—A reservoir that collects and separates liquid and solid wastes, then digests the organic material and passes the liquid waste onto a drainage field or a seepage pit. It is a private system counterpart of a municipal sanitary sewer and treatment plant.

Sewer drain—That part of the drainage system that carries liquid and solid wastes from a dwelling to a sanitary sewer, septic tank, or a cesspool.

Soil stack—A vertical drainpipe that carries wastes toward the sewer drain. The main *soil stack* is the largest vertical drain line of a building into which liquid and solid wastes from branch drains flow. See also *vent stack.*

Soldering—A technique used to produce watertight joints between various types of metal pipes and fittings. Solder, when reduced to molten form by heat, fills the void between two metal surfaces and joins them together.

Solvent-welding—A technique used to produce watertight joints between plastic pipes and fittings. Chemical "cement" softens mating surfaces temporarily and enables them to meld into one.

Stop valve—A device installed in a water supply line, usually near a fixture, that lets you shut off the water supply to one fixture without interrupting service to the rest of the system.

Storm sewer—An underground drainage network designed to collect and carry away water coming into it from storm drains. See also *sanitary sewer.*

Tailpiece—That part of a fixture drain that bridges the gap between the drain outlet and the trap.

Tee—A T-shaped fitting used to tap into a length of pipe at a 90-degree angle for the purposes of beginning a branch line.

Transition—See *adapter.*

Trap—The part of a fixture drain that creates a water seal to prevent sewer gases from penetrating a home's interior. Codes require that all fixtures be trapped.

Uniform Plumbing Code—A nationally recognized set of guidelines prescribing safe plumbing practices. Local codes take precedence over it when the two differ.

Union—A fitting used in runs of threaded pipe to facilitate disconnecting the line (without ever having to cut it).

Vent—The vertical or sloping horizontal portion of a drain line that permits sewer gases to rise out of the house. Every fixture in a house must be vented.

Vent stack—The upper portion of a vertical drain line through which gases pass directly to the outside. The *main vent stack* is the portion of the main vertical drain line above the highest fixture connected to it through which sewer gases from various fixtures escape upward and to the outside.

Water closet—Another name for a toilet.

Water hammer—A loud noise caused by a sudden stop in the flow of water, which causes pipes to repeatedly hit up against a nearby framing member.

Water supply system—The network of pipes and fittings that transports water under pressure to fixtures and other water-using equipment and appliances.

Wet wall—A strategically placed cavity (usually a 2×6 wall) in which the main drain/vent stack and a cluster of supply and drain-waste-vent lines are housed.

Wye—A Y-shaped drainage fitting that serves as the starting point for a branch drain supplying one or more fixtures.

Index

Page numbers in *italics* refer to illustrations.